From Our Galaxy to Beyond: The Story of Cosmic Rays

Sanjay

TABLE OF CONTENTS

Chapter 1

INTRODUCTION

1.1 Context and summary of the book

Some astrophysical phenomena, such as gravitational waves and black holes, were predicted before they were first o bserved. A great m any o thers—from p ulsars to planets to the stars themselves—were observed first a nd r equired a n explanation. Cosmic rays fall in this latter category. Cosmic rays are charged subatomic particles accelerated to ultra-relativistic energies. Relativistic electrons, positrons, and atomic nuclei are all referred to as cosmic rays, but for the most part this book will focus on the hadronic cosmic rays (protons and nuclei), since they reach the highest energies. They are are messengers from some of the most extreme high-energy astrophysical environments, but eleven decades after their discovery, cosmic rays still require an explanation.

The highest-energy cosmic rays likely have extragalactic origins, while less-energetic cosmic rays originate within our Galaxy, but the energy at which this transition occurs and the nature of the highest-energy particle accelerators in the Milky Way are unknown. Observational evidence accumulated over recent decades suggest that a shift from Milky Way cosmic ray sources to extragalactic accelerators occurs for cosmic ray energies somewhere between 1–1000 PeV (see section 1.3). Placing the shift at the lower end of this range creates challenges explaining certain features of the cosmic ray spectrum at high energies, but placing the shift at the high-energy end of the range requires multiple types of Milky Way cosmic ray sources.

Many acceleration models predict that cosmic ray sources produce populations with mass-dependent cutoff energies; hence measurements of the mass composition of the cosmic ray spectrum can disentangle multiple source classes. Recent composition measurements made with the Pierre Auger Observatory conflict with measurements made by Utah's Telescope Array and the LOFAR array (see [30] for comparison of all three). This book describes the process of building a new cosmic ray instrument with the Owens Valley Radio Observatory Long Wavelength Array (OVRO-LWA), designed to help address this tension and search for shifts in composition that may better elucidate the energy limits of Milky Way particle accelerators.

The atmosphere is opaque to cosmic rays, but the flux of high-energy cosmic rays is

too low to be observed directly above the atmosphere with detectors small enough
to launch into space. Thus, these cosmic rays must be observed from the ground, by
observing cascades of energetic particles (air showers) that results from a collision of
a cosmic ray with the atmosphere. One approach to observing cosmic ray air showers
is by measuring their radio emission. The pattern of radio emission observed by an
array of antennas can be used to infer the column depth in the atmosphere at which
the air shower reached a maximum particle density, and this depth is proportional
(on average) to the log of the ratio of energy to the atomic mass number of the
primary particle:

$$X_{\mathrm{max}} \propto ln(E/A)$$

.

Previous radio experiments (see section 1.5) have relied on detecting additional
components of the air shower (e.g. muons, electrons, positrons) in order either to
trigger recording the radio data, or for interpreting the radio data to reconstruct key
shower properties such as energy and composition, or both. For more than fifty
years standalone radio cosmic ray detectors have been a wish [39], but only now
do modern signal processing techniques make such a system possible, by making
the necessary timing accuracy and radio-frequency-interference rejection a tractable
problem. The OVRO-LWA is designed to be the first standalone radio cosmic ray
detector that only requires radio signals for all the detection, RFI rejection, and
analysis steps. Just prior to the upgrade to the OVRO-LWA, Monroe et al. [65]
demonstrated the potential for radio-only cosmic ray detection at the OVRO-LWA.
This book develops a cosmic ray detection system for the upgraded OVRO-LWA.

Chapter 2 describes the major upgrade to the OVRO-LWA that enabled the addition
of a cosmic ray detector, with an emphasis on the aspects that are most relevant to
the cosmic ray system. In particular, digitizing signals from all antennas in a central
location enables the cosmic ray radio-only trigger strategy, in which a coincident
event search implemented in field programmable gate arrays (FPGAs) uses distant
antennas to veto RFI, and saves data from a 20-microsecond buffer for every antenna
whenever a subset of antennas meets the trigger condition. In Chapter 3, I describe
my design and implementation of the FPGA firmware for the cosmic ray radio-only
trigger. In Chapter 4 I describe the commissioning tests for this system.

Chapter 5 returns to the overarching question of *What are the Milky Way's high-
energy particle accelerators?* by focusing on one under-explored Galactic reservoir
of relativistic particles: the magnetosphere of a low-mass flare star. This is the only

part of the book which deals with cosmic ray electrons, although I do explore the prospects of searching for high-energy hadron acceleration in these objects.

The remainder of this expository chapter introduces the context and background necessary to understand this work. Section 1.2 introduces a few key considerations in cosmic ray astronomy. Section 1.3 discusses the evidence for a high-energy limit to the Milky Way's particle accelerators. Section 1.4 introduces ways of measuring cosmic ray composition. Section 1.5 summarizes radio observations of cosmic ray air showers. Section 1.6 takes a broader view of cosmic ray science in our current era of vibrant enthusiasm for multi-messenger astronomy.

1.2 Cosmic Rays

No single class of accelerators can explain the observed spectrum of cosmic rays over the ten orders of magnitude in energy for which we observe a nearly constant power law of flux (Figure 1.1).

The fact that charged particles travel on curved trajectories in magnetic fields leads to both the chief difficulty of cosmic ray astronomy and an important clue to assessing candidates for their origin.

A particle charge q, and momentum $\mathbf{p}$ traveling in a magnetic field $\mathbf{B}$ will gyrate with a turning radius known as the Larmor radius, equal to

$$r = \frac{pc}{qB}$$

where $B = |\mathbf{B}|$, p is the component of the momentum perpendicular to $\mathbf{B}$.

This relation makes it useful to define the *rigidity* of cosmic ray, $R = \frac{c|\mathbf{p}|}{q}$. Substituting the definition of the rigidity into the equation above gives

$$r = R/B$$

and thus the Larmor radius for a particle traveling perpendicular to a magnetic field is the the ratio of the rigidity of the particle to the magnetic field strength in its surroundings.

For ultrarelativistic particles with energy E, $pc \sim E$ and so $R = E/q$.

Disordered magnetic fields in the interstellar and intergalactic media scatter the trajectories of charged particles. Since the radius that a charged particle turns in a magnetic field (Larmor radius) depends on its charge and momentum, the interstellar magnetic field scatters-less easily the higher-energy particles among cosmic rays of the same elemental composition and the lighter nuclei among cosmic rays of the same energy. Particles with Larmor radii comparable to the size of interstellar magnetic field fluctuations travel diffusively. Cosmic ray trajectories do not point back to their sources (except potentially at the highest energies e.g. [44]), which is the reason their origins are so difficult to identify.

The concept of the rigidity also leads to an important constraint on cosmic ray sources. An accelerator must be able to contain the particles that it is accelerating, and so source size and magnetic field strength limit the maximum-rigidity particle a source object could produce. A Hillas plot (see Figure 1.2) compares the maximum-rigidity particles that different astrophysical objects could hold. Neutron stars, white dwarfs, AGN, GRBs, and galaxy clusters satisfy the Hillas criterion up to the highest-energy cosmic rays known, but energy losses to synchrotron radiation may prevent compact accelerators with strong magnetic fields (e.g. neutron stars, white dwarfs, CERN LHC) from reaching their Hillas limit.

Although magnetic fields of course do no work to accelerate the particle, linear accelerators that accelerate cosmic rays to the highest energies observed via a single large drop in electric potential are difficult to arrange in realistic astrophysical environments. Instead, realistic acceleration mechanisms require magnetic fields that confine the cosmic rays to a region in which the particles make multiple passes through a smaller acceleration, gradually reaching high energies. In the scenario of diffusive shock acceleration (e.g. [94] and references therein), repeated crossings of shock fronts in a diffuse gas can produce cosmic rays of the required energies with a spectrum with a power law index of 2.5 which is appealingly close to the observed 2.7 − 3.0. Combinations of acceleration mechanisms may be in effect, such as an initial acceleration over a moderate drop in electric potential followed by a gradual diffusive shock acceleration process. Cosmic rays in the interstellar medium may even amplify the turbulent magnetic fields in which they propagate ([94] and references therein).

The timescale for a cosmic ray in interstellar magnetic fields to lose energy restricts the distance it can travel, and thus the lower-energy cosmic rays, which Milky Way magnetic fields easily confine, must originate within our Galaxy. Supernovae have long been theorised to accelerate cosmic rays [19], and several lines of reasoning favor supernova remnants as Milky Way cosmic ray accelerators. They have suitable turbulent magnetic fields in shocked gas for particle acceleration, and an appropriate energy budget to account for the observed cosmic ray flux. Simulations of supernova remnants produce cosmic rays with energies up to roughly a PeV (or 10 PeV in special circumstellar environments) (e.g. [22] and references therein). Recent ground- and space- based gamma-ray observations of supernova remnants finds spectra consistent with the expected gamma-ray production for collisions of PeV cosmic rays with gas (e.g. [38]).

1.3 The limits of the Milky Way's high-energy particle accelerators: Observational Evidence for a Galactic to Extragalactic shift of the cosmic ray spectrum

Several pieces of observational evidence (see Figure 1.3) suggest the transition occurs somewhere between a few tens and a few thousands of PeV: (1) gradual breaks in the cosmic ray energy spectrum around 1 and 100 PeV and a flattening at a few thousand PeV (called the knee, the second knee, and the ankle, respectively), (2) a shift in anisotropy around 8000 PeV, pointing away from the Galactic plane [2, 85], and (3) several shifts in composition across these energies[1, 7, 30].

If the maximum energy an atomic nucleus with charge Ze achieves in an accelerator object is proportional to its charge (as discussed above), then a single accelerator would be expected to produce a spectrum of cosmic rays with a series of cutoffs at the energies corresponding to different elements' Larmor radii reaching the characteristic size of the accelerator [75]. Since heavier elements have larger Z, they would escape a given source at higher energies than the light elements do.

The gradual bend in the spectrum at the knee, with the steepest part nearly twenty-six times higher energy than the initial steepening could be consistent with protons (mass number A=1) reaching their cutoff energy a factor of twenty-six below the cutoff for iron (A=26) (see e.g. [22] and references therein). The knee does coincide with a

---

[1]Reprinted from Physics Reports, Vol. 872, Julia Becker Tjus and Lukas Merten, *Closing in on the origin of Galactic cosmic rays using multimessenger information*, Copyright (2020), with permission from Elsevier."

[2]Reprinted from Physics Reports, Vol. 872, Julia Becker Tjus and Lukas Merten, *Closing in on the origin of Galactic cosmic rays using multimessenger information*, Copyright (2020), with permission from Elsevier."

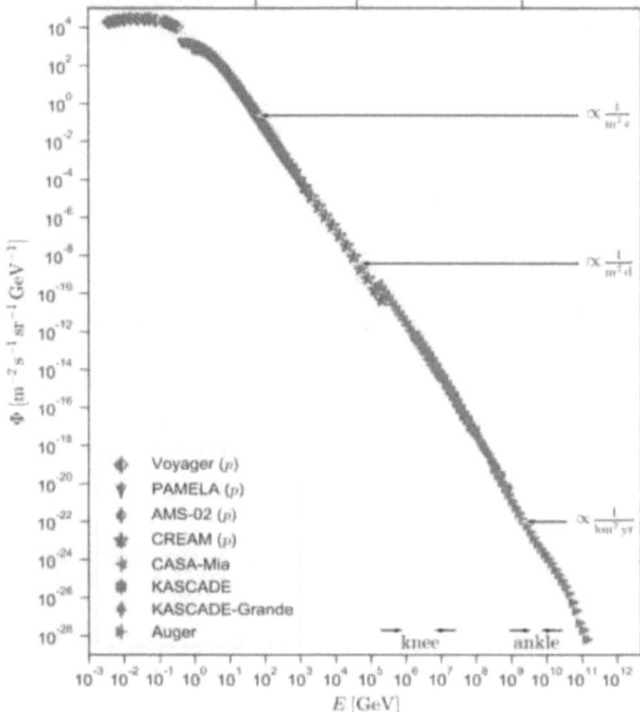

Figure 1.1: Cosmic ray spectrum combining flux observed by various experiments. Red indicates space-based detectors and blue indicates ground-based detectors. Figure reproduced from [22] with permission from the publisher[1]. For the ground-based measurements, the total flux is shown whereas the space-based measurements show the proton-only flux. The discontinuity between the spectrum measured by Voyager and PAMELA satellites is due to the modulation of the cosmic ray flux by the solar wind; the Voyager satellites near the heliopause experience a different flux than does the PAMELA satellite near Earth. The apparent discontinuities at higher energies, however, are primarily due to the systematic uncertainties of the different experiments. Statistical uncertainties are indicated by error bars where larger than the marker size.

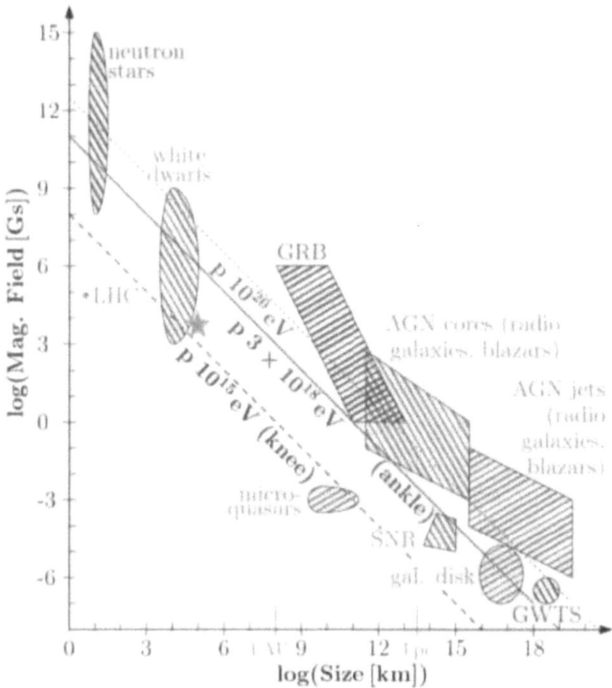

Figure 1.2: Hillas Plot showing the particle rigidities different objects could contain, adapted from [22] with permission from the publisher [2]. I have added the star marker which corresponds to the magnetic field and size of the M dwarf flare star UV Ceti (see Chapter 5).

shift to a heavier elemental composition, although this scenario does not explain the second knee at 100 PeV and the ankle at a few 1000 PeV. Alternatively, the second knee may be the energy at which Galactic accelerators reach their iron cutoff, which would require explaining the ankle as a feature of the extragalactic sources or a propagation effect (e.g. [55]). Other authors, e.g. Thoudam et al. [92], argue for the transition at the ankle, which coincides with the second shift to heavier composition.

Either scenario—a transition near the second knee or a transition near the ankle—has puzzling implications. If the Galactic to extragalactic transition occurs at only a few hundred PeV, diffusion lengths suggest that these extragalactic sources must be relatively nearby, within a few Mpc [70]. If the transition occurs near the ankle, the continued presence of light elements across the second knee [30, 25] requires at least two distinct types of source accelerator within the Milky Way, e.g. [92, 25]. Galactic high-energy cosmic ray sources other than supernova remnants have not been identified, although candidates including X Ray binaries and pulsar wind nebulae have been discussed (e.g. [22] and references therein).

1.4 Cosmic Ray composition measurements from extensive air showers

Observing any particle from any astrophysical phenomena requires that it interact with a material in a way that can be recorded. Human information processing ultimately couples to the universe electromagnetically, and so modern astrophysical instruments fundamentally require that the phenomenon to be observed causes a change in voltage in the instrument. The composition of low energy cosmic rays can be measured to the exact isotope using solid state detectors launched above earth's atmosphere. At energies beyond around 1 PeV the flux of cosmic rays is too low to intercept these small instruments. Detecting cosmic rays at higher energies requires observing their interaction with a larger detection material, such as the Earth's atmosphere.

A cosmic ray colliding with the atmosphere produces a cascade of high-energy particles. The hadronic interactions (including the initial interaction between the cosmic ray nucleus and an air molecule as well as interactions between the air and some of the products of the first collision) produce a large quantity of pions, which typically decay long before they could reach the ground. The charged pions decay to neutrinos and muons, which often reach the ground sooner than they decay into electrons and more neutrinos. The neutral pions decay to gamma rays, some of

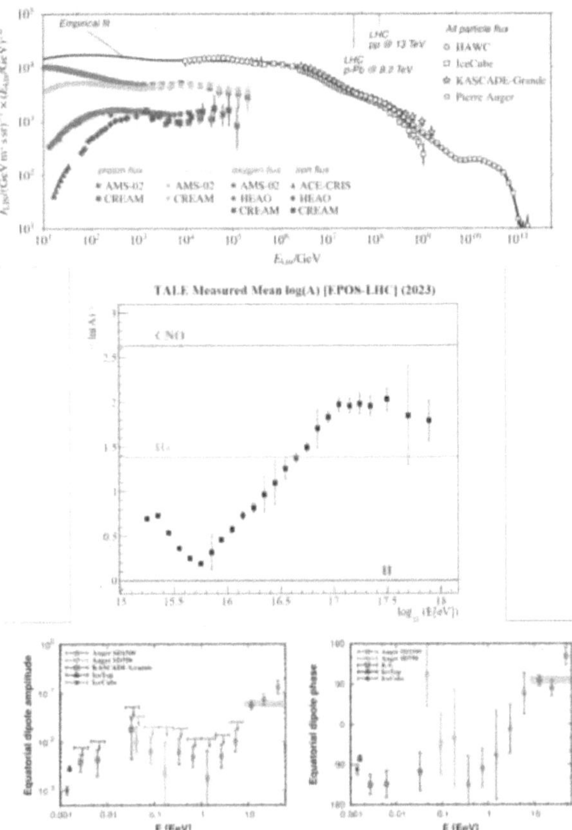

Figure 1.3: Summary of evidence for a galactic to extragalactic transition in cosmic ray sources somewhere between 1–100 PeV. **Top**: Breaks in the spectrum of total cosmic ray flux known as the knee (1 PeV), second knee (100 PeV), and ankle (1000 PeV). The spectrum has been multiplied by $E^{-2.6}$ to highlight these subtle breaks in the overall power law. The colored datasets show the flux for specific elements at energies for which these measurements can be made directly. Figure reproduced from [11] (distributed under a CC-BY license), who adapted it with permission from [32] (distributed under a CC-BY-NC-ND copyright license). **Middle**: Change in composition vs energy (shown as mean log atomic mass number) as measured by TALE (Figure from [8], distributed under CC-BY-NC-ND 4.0 copyright license). **Lower**: Shifts in anisotropy with energy. Figure from [85], distributed under CC-BY-NC-ND 4.0 copyright license.

which pair produce electrons and positrons. Neutrino detection is too difficult to make these neutrinos a useful way to study cosmic rays, although they do create an atmospheric background for dedicated neutrino experiments. The muons, and a portion of the electrons and positrons can be detected with particle detectors placed on the ground (such as solid state scintillators or water Cherenkov instruments). The air shower can also be observed from associated photons, including via Cherenkov radiation, fluorescence of air molecules excited by the particle shower, and radio emission (which will be discussed in detail in section 1.5). Figure 1.4 summarizes the methods of observing the different types of signatures of extensive air showers.

Current ground-based methods cannot measure the atomic mass number of individual cosmic rays, but can estimate overall statistics binned by energy and mass-groups. Several features of an air shower depend on the nature of the primary cosmic ray, and the most readily-observable of these are the distribution of leptons at the ground and the column depth in the atmosphere at which the shower reaches a maximum particle density. This latter quantity is known as X_{max} and varies with the log of the atomic mass number, on average. Since each individual air shower includes a random process of particle collisions, X_{max} has some variance among showers for cosmic rays of equal mass and energy. Thus, individual X_{max} measurements do not identify the mass of individual particles, but the distribution of X_{max} for a sample of cosmic rays describes the relative proportion of heavy and light nuclei (e.g. [6], [30]).

Combined analysis [32] of data from Kascade-Grande (Germany) at 3-1000 PeV, and up to 10^5 PeV from the Pierre Auger Observatory (Argentina) and Telescope Array (Utah) suggest an increase in the heavy-element fraction at the first knee, changing back to light elements after the second knee, with another shift to heavy elements beyond 1000 PeV. However, in the 100-1000 PeV range, the LOFAR array finds a somewhat heavier composition than Auger [30, 25]. At these energies, LOFAR in the Northern hemisphere and Auger in the South are not expected to measure different cosmic ray populations since the Galactic magnetic field should make cosmic rays from all sources nearly isotropic at these energies. However, an explanation arising from a systematic uncertainty in the calibration of one or both instruments has not yet been identified, motivating additional experiments, including this book.

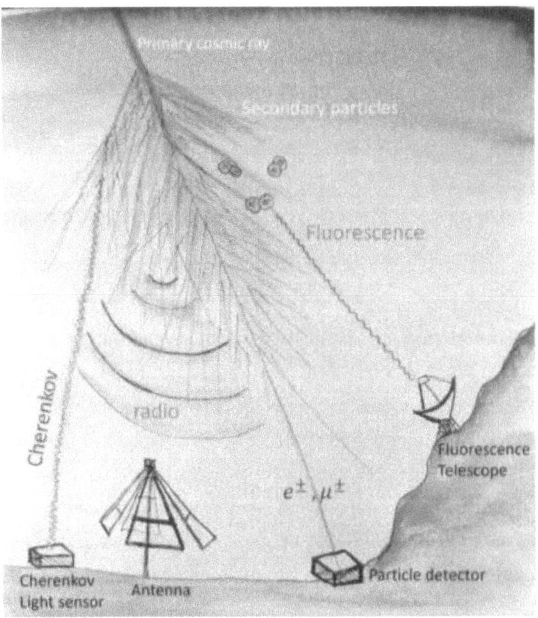

Figure 1.4: Illustration summarizing the different observable signatures of cosmic ray air showers, and their detectors.

1.5 Radio observations of cosmic ray air showers

Origin of radio emission in cosmic ray air showers

Radio emission in extensive air showers mainly originates by two mechanisms (see e.g. [14]). The electrons and positrons in the shower radiate coherently in the Earth's magnetic field, creating radio emission referred to as the geomagnetic emission. The geomagnetic emission results in a brief increase in the electric field in the $\mathbf{v} \times \mathbf{B}$ direction, where $\mathbf{v}$ is the direction of shower propagation and $\mathbf{B}$ is the direction of the Earth's magnetic field. Additionally, a charge excess accumulates at the front of the shower due to a combination of effects. The radiation produced by these charges is Askaryan radiation [17], and although this mechanism was the initial radio emission mechanism of interest after the determination that observing radio Cherenkov emission was unlikely (see [39] and references therein), it constitutes a much smaller contribution to the radio emission from air showers but becomes more significant for dense media such as rock and ice.

The sum result of the main contribution from geomagnetic emission and a smaller contribution from Askaryan radiation is a pulse in the electric field on roughly 1–10 nanosecond timescales depending on the position of the viewer relative to the shower axis. This radiation is beamed and illuminates an area roughly the size of the Cherenkov ring (see e.g. [14]). The amplitude of the electric field of the radio pulse is approximately linearly proportional to the energy of the shower. Figure 1.5 shows the radio emission for a simulated air shower. This emission is best observed from 10–100 MHz, and within that window 30–80 MHz is the easiest band to observe due to having less RFI.

The most precise numerical simulations of radio emission from air showers calculate the radio emission as the superposition of contributions from instantaneous accelerations of shower particles along their trajectories, taking the populations of shower particles from a separate simulation of the hadronic interactions in the cascade [13], reviewed in [14]). The state-of-the-art method for estimating X_{max} from air showers measured with radio antennas consists of forward modelling showers with different X_{max} and energies, computing the radio emission from each of these modelled showers, and then determining which simulation best matches the observed radio signal (e.g. [30, 27]).

The hadrons in cosmic ray air showers interact at much higher energies than can be studied directly in accelerator laboratories, and at these energies the several leading hadronic shower simulations have systematic offsets among them in their predicted mean and variance of X_{max} for each element (see e.g. [76]) by up to $25 g/cm^2$ in the 100-1000 PeV energy range of interest to this book. One advantage of observing cosmic rays by their radio emission is that electrodynamics is understood so much better than the strong force. For a given radio observation of a cosmic ray air shower, the estimated X_{max} is not sensitive to the systematic uncertainty between hadronic models. Observed shifts in X_{max} as a function of energy can be interpreted as shifts in composition to heavier or lighter elements independently of attempting to identify the exact average atomic mass number of the population.

Past, Present, and Near-future Radio Experiments
JELLEY et al. [53] made the first observation of radio emission from cosmic ray air showers in 1965 by comparing 44 MHz radio signals to Geiger counter counts. Additional experiments in the 1960s and early 1970s confirmed important features of the emission, including the importance of the geomagnetic emission mechanism

(reviewed in e.g. [39]). Standalone radio systems were a goal in this early period, in large part due to the lower cost and relative simplicity of radio antennas compared to particle counters, but radio frequency interference and the requisite timing precision for comparing data among different antennas meant that radio experiments continued to depend on particle arrays [39]. Later in the 1970s, the field largely abandoned radio observations of cosmic rays, partly due to the challenges of a standalone radio system [39].

Modern radio cosmic ray experiments tend to either be built at existing astroparticle observatories (e.g. Tunka Rex, Lopes, AERA, ICE TOP) or at existing radio astronomy observatories (LOFAR, MWA, OVRO-LWA, SKA, CODALEMA). Renewed interest in radio observations of particle showers came when modern digital signal processing made made possible the successful detection of cosmic ray air showers with the LOPES[37] and CODALEMA [81] experiments. LOPES used trigger signals from the co-located Kascade particle detector array [37] and CODALEMA used a combination of triggers from particle scintillators and radio self-triggers with post-data-recording cross comparison to the scintillator array. Furthermore, prospects of detecting high-energy astrophysical neutrinos via the Askaryan radiation of particle showers in dense media, such as ice, sparked additional interest in radio techniques for observing extensive particle showers (see e.g. [14] and references therein).

Cosmic ray detection at the LOFAR array directly followed from the LOPES experiment, with radio data recorded on receipt of trigger signals from the LORA particle scintillator array (see e.g. [25]). The size of the LOFAR superterp array makes LOFAR best suited to $10^{16.5}$–10^{17} eV cosmic rays, similar to the OVRO-LWA and Tunka-Rex, although Tunka-Rex is a much sparser array, with 20 antennas in a roughly square km area [88]. At Tunka-Rex, radio signals from the antennas are digitized by the same ADC cards as serve the photomultiplier tubes of the larger Tunka non-imaging Cherenkov detector array, allowing cross-calibration between the Cherenkov and radio signals [88]. The AERA array at the Pierre Auger observatory consists of 153 antennas over a 27 square km area, with most antennas triggering data recording from signals from the surface particle detectors and a smaller number self-triggering on radio data alone (see e.g. [79]. However, X_{max} estimates are only made for events that have a trigger from the particle array [79]. AERA and LOFAR have showed that state-of-the-art X_{max} reconstruction is possible with the radio technique [30, 79].

New radio cosmic ray detection experiments are underway at the Murchison Wide-

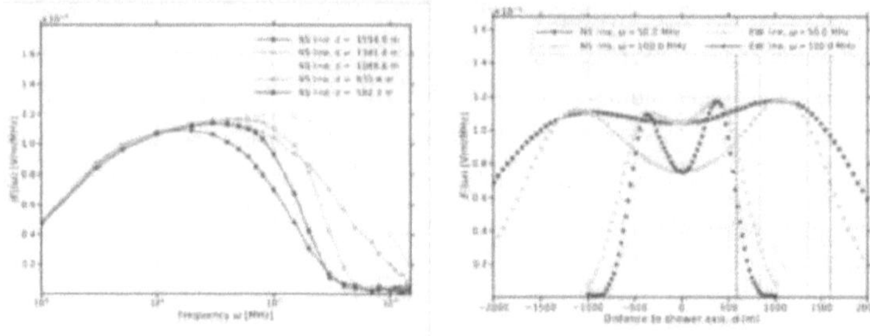

Figure 1.5: Radio emission from a simulated 1000 PeV proton arriving from the North with a zenith angle of 70 degrees. Left: Electric field spectrum at different positions along a North-South line through the shower. Right: Peak electric field spectral density as a function of distance from the shower axis, plotted for different frequencies and lines through the shower. Figure adapted from [14] (distributed under a CC-BY license).

field aray [98] and the ICE-TOP array at the ICE-CUBE neutrino observatory [93], and these both reference radio signals to particle scintillators. The future Square Kilometer Array plans to include a cosmic ray search, and the dense antenna spacings and large extent will make this array extremely good at reconstructing cosmic ray air showers [51].

The OVRO-LWA is designed to be the first array to use radio signals alone, without a particle scintillator trigger, for triggering data readout, flagging RFI, and reconstructing key shower observables (energy, X_{max} , arrival direction). Toward these goals, this book work implements an event search and RFI rejection strategy on the field programmable gate array boards (see Chapters 2 and 3) that process the raw ADC output. An advantage of using radio signals alone is avoiding the complicated selection effects of hybrid radio and scintillator arrays (see e.g.[30]).

An additional advantage of the OVRO-LWA is that its cosmic ray detector will observe simultaneously with other astronomy observations. Simultaneous observing will accumulate a large sample of cosmic rays much faster than observing in limited, scheduled sessions. The OVRO-LWA is expected to detect 2000 cosmic rays per year, and simulations by [27] predict X_{max} reconstruction with a state-of-the-art precision better than $20 \, \mathrm{g/cm^2}$.

1.6 Cosmic rays in the landscape of multimessenger astronomy

Studying any astronomical phenomena requires that a particle involved in the phenomenon reach Earth, or cause another particle to reach Earth. For thousands of years, astronomy relied on photons as the messengers. Just over a century ago cosmic rays began to be observed, followed by neutrinos decades later, and finally gravitational waves within the most recent decade. Combinations of messengers provide a richer picture of the universe.

The Astro2020 Decadal Survey [69] identified time domain and multi-messenger astronomy as a priority for astrophysics in the next decade. Conclusively identifying the high-energy cosmic ray accelerators is a multimessenger endeavor. Cosmic ray interactions with gas in and around the accelerator objects or in the interstellar medium should produce gamma-rays and neutrinos, and these neutral messengers travel undeflected by magnetic fields on trajectories that, unlike cosmic rays, do point back to their sources. Recent observations of high-energy gamma-ray emission from supernova remnants, massive star forming regions, and the Galactic center highlight intriguing possibilities for Milky Way particle accelerators (e.g. [38, 10, 9]). These gamma-rays could, however, emanate entirely from leptonic processes—such as inverse Compton scattering off cosmic ray electrons—and do not definitively identify these objects as the elusive sources of hadronic cosmic rays. Complicating these multimessenger searches, not all source models would expect cosmic rays to interact with other particles before escaping, and conversely hadronic accelerators surrounded by very dense regions could prevent the escape of gamma rays and possibly the cosmic rays as well such that only neutrinos emerge [91]. A few recent gamma ray observations (e.g. [38]) detected a feature ascribed to pion decay which will be strong evidence of cosmic ray acceleration if supported by future observations. Radio observations are key to constraining populations of cosmic ray electrons in these sources.

Compared to gamma ray emission alone, identifying high-energy neutrinos to an astrophysical object is a much more likely sign—but perhaps not actually a 'smoking gun' (Hooper and Plant, 2023 submitted)—that the object accelerates hadronic cosmic rays. Thus far, astrophysical neutrinos have been identified with the AGN of the galaxy NGC1068 [4]), as well as a tentative association with the blazar TXS 0506+056 [3]. No specific Milky Way sources have been identified, however, although an over-density is observed along the Galactic plane [5].

Upcoming instruments and observations have exciting potential to disentangle some of these puzzles, but due to the complexities of interpreting potential signatures of cosmic ray interactions, better measurements of the cosmic rays themselves will complement these efforts.

Chapter 2

INCORPORATING A COSMIC RAY SEARCH IN THE UPGRADE TO THE OVRO-LWA

2.1 Introduction

Caltech's Owens Valley Radio observatory is in a sparsely-populated region of Eastern California, in a narrow valley between two tall mountain ranges that each exceed 4000 meters above sea level. The Sierra Nevada to the west and the White Mountains to the east help block radio frequency interference from major cities in this westernmost valley of the north American Great Basin basin-and-range geology. The observatory at the valley floor sits at an elevation of 1200 meters, well below X_{max} for the air showers we expect to observe. A 100 PeV proton arriving from zenith with a typical X_{max} of 700 g/cm^2 will reach X_{max} at an altitude of roughly 3000 meters.

The OVRO-LWA is an array of dual polarization dipoles built to study a wide range of phenomena including extrasolar space weather, the cosmic dawn epoch, solar flares, as well as now searching for high-energy cosmic rays. The OVRO-LWA was built in three stages. The first stage consisted of 256 dual-polarization dipole antennas in a dense array roughly 200 m in diameter. A second stage added 32 antennas outside this dense core array [36, 15]. Finally, from 2019 to 2023 the array underwent a 2.6 million dollar NSF-funded upgrade that consisted of expanding from 288 to 352 antennas within a 2 km extent, re-designing the analog electronics for better signal path isolation and better sensitivity to lower frequencies, and entirely replacing the digital electronics. Just prior to the upgrade, Monroe et al. [65] demonstrated the array's potential as a cosmic ray observatory by detecting eight cosmic ray air showers in a dedicated 40 hour observation. The new digital signal processing electronics in the upgrade created the opportunity to build a cosmic ray detection system that will search for air showers nearly-continuously, alongside the other types of astronomy observation.

This chapter describes the upgrade to the OVRO-LWA, with an emphasis on aspects that are relevant to the cosmic ray detection system. Section 2.2 describes the array. Section 2.3 provides an overview of the cosmic ray detection strategy. Section 2.4 describes the timing and synchronization strategy. Section 2.5 describes the development of a means of communicating fast cosmic ray trigger signals between SNAP2 boards. Section 2.6 explains the system for transmitting signals from the most distant antennas back to the central signal processing shelter. Relevant to this chapter, appendix A provides a manual for the set up of the SNAP2 field programmable gate array boards that process the output of the analog to digital converters. Section 2.7 presents simple simulations comparing several array layouts from the perspective of optimizing cosmic ray detection. Section 2.8 describes an investigation on the effects of vegetation on antenna response to RFI, and section 2.9 concludes the chapter.

2.2 The OVRO-LWA

Figure 2.1 shows an LWA antenna as well as a typical power spectrum observed by both polarizations of a typical antenna. The antenna front-end electronics include a low noise amplifier. The antennas' sensitivity is dominated by Galactic synchrotron background over two octaves in frequency, from 18 to 85 MHz. Each antenna stand has one dipole aligned to geographic North-South and the other dipole aligned to geographic East-West. These antennas are not designed to distinguish horizontal and vertical polarization. In writing this book, there are contexts where 'antenna' refers to a dual-polarization pair of dipoles, as well as contexts (such as within the firmware) where 'antenna' refers to a single dipole. I will specify whether I'm referring to individual dipoles or dual-polarization pairs as needed.

The layout of the OVRO-LWA is optimized for astronomical imaging and consists of irregularly-spaced antennas in a dense core of 200 meters in diameter, surrounded by sparsely-spaced antennas out to a maximum separation of 2.4 km. Most cosmic rays will be detected within the core of the array, although the sparser regions will be sensitive to inclined showers at higher energies. Figure 2.2 shows the array layout with radio emission from a simulated cosmic ray air shower overlaid.

Signals from all antennas are continuously digitally sampled in a centrally-located electronics shelter (see Figure 2.3) which contains the computing cluster for the astronomy correlator, as well as a dedicated computer for processing cosmic ray data. The analog signals are transmitted to this central electronics shelter using coaxial cable for antennas in the dense core array and RF-modulated fiber optics for the distant antennas. After passing through analog filters and attenuators in analog receiver circuit boards designed by Larry D'Addario, the signals are digitized by 10-bit ADCs at a sample rate of 196 MHz. The analog receiver boards allow several options for filters and attenuation settings, controllable from software. Nearly all observations, including the cosmic ray detector, use the filter option with a high-pass cutoff at 18 MHz and a low-pass cutoff at 85 MHz. The attenuation settings are adjusted for each antenna with the goal of compensating for the different cable attenuations such that all ADC inputs receive similar power levels. A major goal of the re-design of the analog receiver boards is reduction of cross-talk by which signals leak into adjacent channels. The new analog receiver boards are designed to more than 80 dB

Signals of both polarizations from each antenna are digitized with ten-bit analog-to-digital converters with a sample rate of 196 MHz. The full array will produce data at 1.40 terabits per second. The first stages of digital signal processing are performed on 11 field programmable gate array (FPGA) boards, which are the second generation of the SNAP2[1], containing FPGAs, two 1 Gb Ethernet ports, and four 40 Gb QSFP+ Ethernet ports (see Figure 2.4). Each SNAP2 contains two FPGAs: a Xilinx Kintex Ultrascale+ and a smaller Xilinx Zynq. The larger Ultrascale+ performs all the main signal processing, including the filterbank firmware and the cosmic ray search and additionally runs a Microblaze CPU that facilitates control of the FPGA. The smaller Zynq FPGA is used for tasks such as monitoring the SNAP2's power use.

Each SNAP2 processes signals from both polarizations of 32 antennas (four ADC cards designed by Larry D'Addario that handle 16 channels each connect to each SNAP2). These SNAP2 boards run the first stage of the cosmic ray search as well as a filterbank that computes frequency-domain channels for the other observing modes. The next section gives an introduction to the radio only self-trigger strategy. For a detailed description of the cosmic ray trigger firmware see Chapter 3.

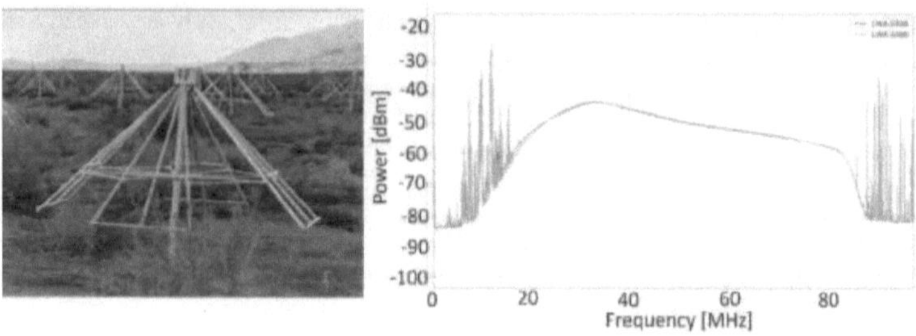

Figure 2.1: Left: An OVRO-LWA antenna, with other antennas in the dense core array behind it. The OVRO-LWA antennas consist of a pair of dipoles shaped to be sky-noise dominated over two octaves of frequencies. Front end electronics sit under the white square cover at the top of the antenna. Right: Typical spectrum observed by an LWA antenna (this spectrum is from the same antenna as pictured, but the photo and the spectra were taken at different times). The orange trace shows the spectrum from the East-West aligned dipole (polarization A) and the blue trace shows the spectrum from the North-South aligned dipole (polarization B). The photo is taken by the author and the spectrum plot is adapted from Greg Hellbourg's RFI monitor tool.

2.3 OVRO-LWA Radio-only cosmic ray self-trigger: An overview

This section presents an overview of the design of the OVRO-LWA cosmic ray detection strategy, in order to enable cosmic ray detection from radio emission alone, without referring to external particle detectors. I include an overview of the trigger firmware, and an introduction to the RFI rejection strategy. The details of the design, testing, and performance of the various subsystems are presented in the later sections of this chapter.

The radio-only self-trigger

The first stage of the search must occur on the FPGAs, because cosmic ray air shower pulses are shorter than the 20 ns impulse response of the OVRO-LWA's analog electronics. In order to detect such fast events, the detection must use the raw ADC output timeseries and take place on the FPGAs. Since the upgraded array replaces all the signal processing electronics, I build on the technique of [65] but write all new firmware and software. The cosmic ray firmware will operate in parallel with the filterbank firmware.

Figure 2.6 outlines the trigger firmware, which is described in detail in Chapter 3. The SNAP2s buffer 20 microseconds of the time series output by the ADCs,

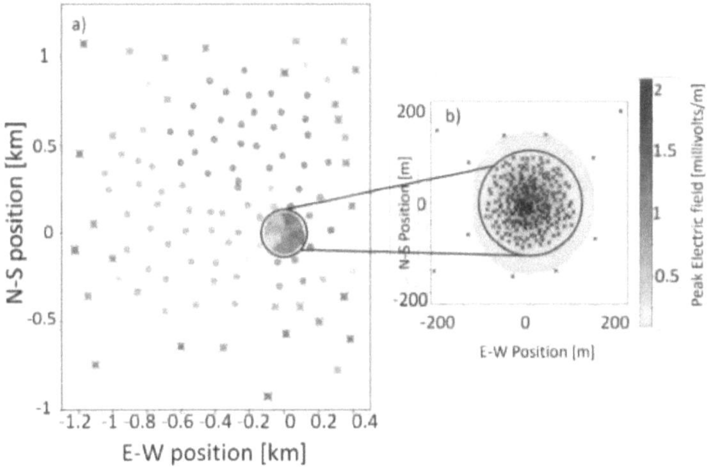

Figure 2.2: Array layout of the OVRO-LWA. A) Circles and Xs mark antenna positions. Xs mark antennas whose role is to veto RFI (see section 2.3). Antennas are color-coded to indicate which groups of antennas are processed by the same FPGA. B) Zoom in on the core array, overlaid with emission from a simulated cosmic ray air shower. Xs mark all antenna positions; none of these antennas are part of the RFI veto. The color indicates the peak electric field from the air shower of a 100 PeV proton arriving from zenith.

with ~ 5-nanosecond time resolution. This buffer uses Block RAM (BRAM) implemented as a circular buffer which continuously overwrites old samples such that it always contains the most recent $20\,\mu s$ of data.

In parallel with the buffer, the time series are filtered and thresholded to search for pulses. Each SNAP2 searches both dipoles for 32 antennas (64 signals total). Within the firmware, every signal is treated as an individual antenna; knowledge of polarization pairs is useful later in software processing. If signals above a power threshold occur from more than a threshold number of antennas, within the light travel time between those antennas, then the detector logic outputs a trigger signal. The trigger can be cancelled by RFI rejection logic that will be described in the next section. If the trigger is not cancelled by the RFI blocker, then the board stops writing new data to the buffer and transmits the buffered timeseries over Ethernet to a computer for the next stages of processing.

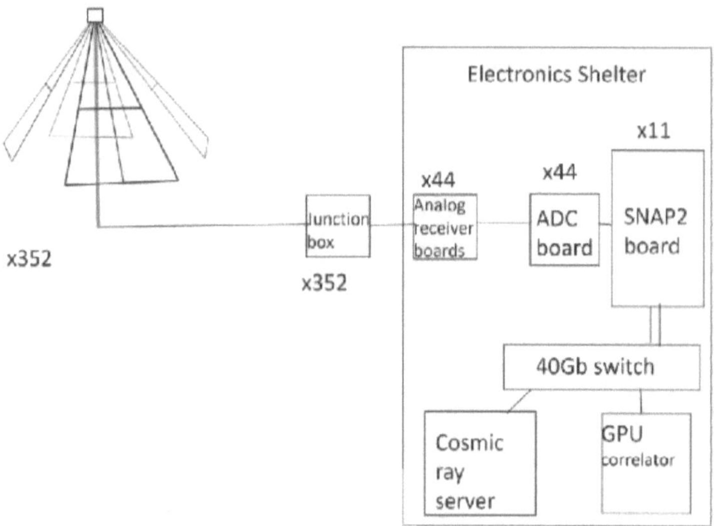

Figure 2.3: Simplified diagram of the OVRO-LWA signal path. Signals from 352 dual polarization antennas are transmitted to a central location, where 44 analog circuit board filter 16 signals each before the signals are digitized by 44 ADC boards which sample 16 signals each. Eleven SNAP2 boards process the output of the ADCs, and the cosmic ray trigger logic runs on these boards. A 40Gb switch allows cosmic ray data to be transmitted to a separate computer from the correlator.

In searching for events, first a time-domain finite impulse response (FIR) bandpass filter is applied to the raw voltage timeseries to suppress RFI in the FM band and below 30 MHz, in addition to the suppression achieved by the analog electronics. Then, the firmware squares the voltages to produce a power timeseries, and smooths the power timeseries with a 4-tap moving window average. The pulse detector logic outputs a Boolean 1 when the smoothed power signal exceeds a threshold controlled from software. The firmware counts the number of antennas that detect the event by extending the Boolean pulses according to the light travel time between the antennas and then summing across antennas. For a cosmic ray to be detected, a threshold number of antennas processed by an individual FPGA must detect the pulse.

If the trigger condition is met for any SNAP2 board, all the SNAP2 boards read out their buffers, so that a snapshot is saved for the entire array, including the antennas whose signal did not meet the trigger condition. This will allow sub-threshold pulses

Figure 2.4: SNAP2 boards set up in the lab for testing before being installed in the OVRO-LWA. Black letters label certain important features: A) 40 Gb QSFP+ Ethernet ports. B) 1 Gb Ethernet ports. C) The Xilinx Kintex Ultrascale+ FPGA is under this heat sink and fan— the fan is used in the lab only, since in the OVRO-LWA cooled air flows through the entire electronics rack. D) One of two FPGA-mezzanine-card (FMC) connectors. Four ADC cards digitizing 16 signals each attach to each board, with a stacked pair of ADC cards attaching to each FMC connector. E) Power supply connector. F) Pin header for programming the board via JTAG. After using JTAG to load an initial firmware configuration, subsequent re-programmings can be acheived using the 1 Gb Ethernet interface.

to be used in the air shower reconstruction. To accomplish this readout from the entire array, all the 11 SNAP2 boards are wired to each other in a loop linking the their General Purpose Input Output Pins (see section 2.5). Whenever the firmware on one SNAP2 board begins reading out its buffer, it transmits a signal to the other 10 SNAP2 boards to ensure that the processing computer receives a snapshot from the entire array. The initial trigger detection must occur within the subset of antennas that an individual FPGA processes, but any FPGA can trigger the rest of the array to read out data. After a 20 μs snapshot is read out over Ethernet, the FPGA resumes writing to the buffer and awaits new triggers.

When a trigger signal halts writing to the buffer, the buffered 20 μs timeseries for each antenna is read out into UDP Ethernet packets and transmitted from a QSFP+ Ethernet port to a dedicated computer via the same 40 Gb Ethernet switch that also serves the astronomy correlator. The readout deadtime is determined by how fast the receiving computer can process the Ethernet packets, not by the maximum speed that the SNAP2 boards can transmit the data (see Chapter 3). Every readout has a detector dead time of 0.7 ms. At a trigger rate of 55 Hz this results in a 4% fractional dead time.

Figure 2.5 outlines the different types of information flow required by the system. Trigger signals will be communicated between FPGAs. Raw voltage data for detected candidates is transmitted from the FPGAs to the CPUs over a 40 Gb Ethernet network. The CPUs transmit system configuration information such as thresholds and cable delays to software registers on the FPGAs over a 1 Gb Ethernet network.

The cosmic ray system's main use of FPGA resources is BRAM for the voltage buffer and DSPs for the FIR filter. Since much of the trigger process uses Boolean signals, resource use is small in this part of the system. The FIR uses roughly half the available DSPs in the current version, and a future, optimized version will seek to reduce this resource use.

Introduction to the Strategy for Rejecting Impulsive RFI

Monroe et al. [65] encountered impulsive RFI at an estimated rate of 500 Hz. This background necessitates building a first stage of impulsive RFI event rejection into

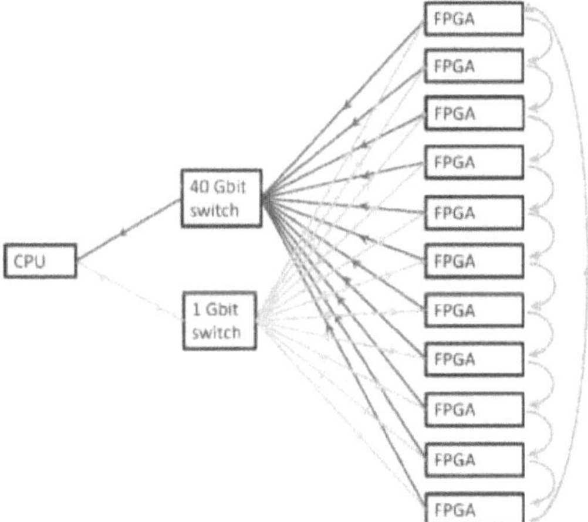

Figure 2.5: Flowchart summarizing the information flow in the cosmic ray detection system. Configuration settings and system monitoring information travel between the FGPAs and a central computer on a 1 Gb Ethernet network (gold arrows). Timeseries data travel from the FGPAs to the central computer on a 40 Gb Ethernet network (purple arrows). Trigger signals travel from each FPGA to its neighbor in a loop on a dedicated wire between general purpose input output pins on the FPGA boards (green arrows).

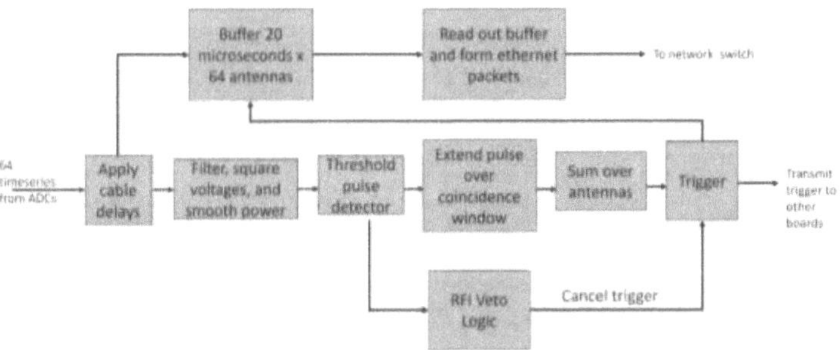

Figure 2.6: Flowchart summarizing the trigger logic for the cosmic ray radio-only trigger. This process is implemented in firmware running on the SNAP2 FGPA boards.

the firmware on the FPGAs, to keep the trigger rate low enough to avoid substantial Ethernet packet loss in the 40 Gb network or at the receiving computer, and to minimize readout dead time. The digital FIR filter (see Chapter 4) rejects persistent RFI below 30 MHz and in the FM band, but broadband impulsive RFI requires a different strategy. Thus, to operate a radio-only self trigger in the presence of radio frequency interference (RFI), a first stage of RFI rejection must occur on the FPGAs. Later stages of RFI rejection occur in software processing on CPUs after the snapshots of data have been transmitted off the FPGAs.

Since cosmic ray air showers are beamed, I use distant antennas to veto RFI within the firmware. Prior to the upgrade, in the 256 element LWA, antennas were grouped to FPGAs in sets of 16 arranged in North-South stripes and only the core array was used for the demonstration cosmic ray search. The upgrade to 352 element array made all of the antennas available for the cosmic ray detector, and also offered the opportunity for the groupings of antennas to FPGAs to be optimized for the cosmic ray detector. Figure 2.2 shows the layout for the upgraded array, color-coded according to which of the 11 SNAP2 FPGA boards processes data from each antenna. The grouping of antennas to FPGAs does not affect any other science purposes of the OVRO-LWA other than the cosmic ray detector, and so the groupings were optimized for cosmic ray detection as follows. For antennas in the core array, groups of neighboring antennas are processed by the same FPGA. This arrangement increases the probability that an air shower will illuminate enough antennas within the same FPGA to meet the trigger condition.

Each FPGA also processes a few distant antennas, separated by a sufficient distance that most cosmic ray air showers in the energy range of interest would not illuminate the distant antennas on both sides of the array. Within each FPGA, when the trigger condition is met in the core antennas, the trigger is canceled if an above-threshold signal is also detected by a threshold number of the distant veto antennas. If the trigger is not vetoed, then the buffered snapshot of data is transmitted off the FPGAs. Importantly, the veto system requires that multiple veto antennas from the same FPGA detect an event in order to reject it, thus requiring detections by veto antennas on opposite sides of the core. Note in Figure 2.2 that some of the antennas marked as veto antennas to the south of the core array are in fact near enough to the core to be illuminated by some air showers that illuminate the core. The FPGAs that use these veto antennas have at least one other veto antenna that is more distant on the opposite side. The antennas on the north side are far enough that only extremely inclined air showers or air showers from cosmic rays with energies well above the second knee would simultaneously illuminate these veto antennas and core or south antennas.

Antennas in the dense core of the array are divided into nine regions that are each processed by one FPGA, so that neighboring antennas are processed by the same FPGA. These groups are arranged to improve the rate at which a single FPGA will have enough antennas within an air shower footprint to detect the air shower. Additionally, each FPGA will process signals from a few distant antennas up to 2.4 km away, which act as an RFI veto. These antennas would lie outside the radio footprint of cosmic rays whose shower axis is near the core for the energies we are interested in, and so radio impulses detected by both the core antennas and the very distant outliers can be rejected as RFI. Figure 2.2 shows the array layout with antennas color-coded by FPGA, with veto antennas marked. Most of the sparse area of the array is grouped to two FPGAs, which are currently not the focus of the cosmic ray search but could eventually be used to have more sensitivity to higher-energy cosmic rays than the core array provides.

Further stages of RFI rejection take place in software, after snapshots of buffered data are transmitted from the whole array to one receiving computer, as discussed in Chapter 4.

2.4   Timing and clock distribution

Searching for radio emission from cosmic ray air showers, and estimating the showers' X_{max} ultimately depends on knowledge of the time evolution of the electric field at different positions on the ground. Comparing the time evolution of the electric field in different positions in space requires that different measurements can be compared to a common time reference. Assigning times to data has two components: relative timing and absolute timing. Relative timing refers to the precision and accuracy with which measurements from different antennas can be compared to each other. Absolute timing refers to the precision and accuracy with which measurements can be compared to time references outside the observatory.

The goal of cosmic ray detection for precise X_{max} measurements places much stronger requirements on relative timing than on absolute timing. Relative timing is crucial because identifying cosmic ray air showers and measuring X_{max} depends on being able to compare the arrival times of electric field impulses at different antennas in different places, to within the order of the nanosecond timescale of the original (not bandlimited) pulse. The initial detection of the cosmic ray air showers within the FPGAs depends on detecting an above-threshold electric field at more than a minimum number of antennas, within the light travel time between those antennas. Since this search is accomplished by counting events within a window, the initial detection places a less strict requirement on timing than the reconstruction does. Making the window slightly larger than the maximum light travel time to account for timing uncertainty does not result in a large increase in the candidate event rate, and so relative timing uncertainty of several tens of nanoseconds is tolerable at the stage of detecting candidate events. The system currently uses a combination of cable delays that were measured for antennas that existed prior to the upgrade and predicted delays for the new antennas, based on the known cable and fiber lengths. The pre-existing measurements used a reflectometry device to measure the delay in the long cable from the antenna to the shelter. The components of the signal path within the shelter are expected to make a much smaller contribution since these components are similar between antennas. In the near future, the electronics delays will be measured more precisely for all antennas.

The absolute timing is even less important, because at the energies of interest, cosmic rays' diffusion through Galactic magnetic fields prevents the cosmic rays' arrival times and directions from revealing compact sources or fast transient events. Absolute timing matters more for the interferometric imaging capabilities than for the cosmic ray system. While good relative timing suffices for operating the correlator, absolute timing precision determines the ability to place images in a common astronomical coordinate system, and supports plans eventually use the OVRO-LWA in a very long baseline interferometer with the New Mexico LWA stations.

There are two main sources of relative timing uncertainty: delays in the analog electronics and knowledge of the times measurements are sampled. Analog to digital conversion occurs in one signal processing shelter for all 704 signals. If an ADC samples the voltage at an input channel at a particular known time, what we want to know is what time did the electric field occur at the antenna that produced that voltage on the ADC input. Thus, the amount of time that the signal takes to propagate through all of the analog electronics—front-end amplifiers, coaxial cables and (if applicable) optical fibers, and the electronics in the analog receiver boards—must be accounted for. I refer to this time as the electronics and cable delays. In fact, since the relative timing is the main concern, only the differences between the signal paths for different antennas matter. Any components that produce an identical delay on all the antennas have no affect on the relative timing. The most significant difference in the electronics and cable delays for different antennas is the different cable and optical fiber lengths between the antennas and the signal processing shelter.

The firmware in the FPGAs that process the ADC output contains a logic block that delays all of the digital signals by a different amount to compensate for the relative differences in cable delays, to the nearest integer number of clock cycles. The values of these delays are written to the FPGAs from software each time the FPGA restarts. As a first approximation, values of the delays are obtained from the cable and optical fiber specifications. More precise delay calibration can be applied to data after it is recorded, by comparing the arrival times of a known source at different antennas. Monroe et al. [65] used known RFI sources for improving the relative timing.

Uncertainties in antenna positions would create errors that resemble cable delays. From the construction process, the positions are known to within roughly a meter. Antenna position measurements could be improved in the future with a survey device that already exists at the observatory which uses GNSS signals for position measurements.

The second major component of the relative timing uncertainty is knowledge of the times the voltages are sampled. This synchronization and timestamping is accomplished as follows (see Figure 2.4). A coaxial cable from the main observatory clock carries a 10 MHz reference signal which includes a 1 Hz signal by missing one pulse every second. A clock device in the LWA signal processing shelter takes that observatory clock signal and outputs a Pulse-Per-Second (PPS) signal and a 10 MHz reference separately. A Valon signal generator locks to the 10 MHz signal and outputs a 196 MHz sinusoid which is split to all of the 44 ADCs cards using 44 cables of the same length. The clock that drives the FPGAs on the SNAP2s is taken from the ADC cards. Each ADC samples signals from 16 antennas, and a combination of ADC hardware and FPGA firmware logic ensures that all 64 data streams within one SNAP2 board are synchronized with respect to each other.

Synchronizing the 44 ADC clock signals, such that on every clock cycle 704 antenna signals (counting both polarizations of 352 antennas) are sampled simultaneously does not fully address the timing requirements. An additional system is required to allow comparing data between boards, by labelling the data with a timestamp. The PPS signal is sufficiently accurate to time-tag data to the correct second, but may not have a steep enough rising edge to resolve the 196 MHz clock cycles. In other words, there may not be exactly 196 million clock cycles between successive pulses [31]. Thus, the PPS signal is input to one SNAP2 board, which distributes a single-clock cycle synchronization pulse to the other 10 boards and back to itself (via a twelve-way splitter with one unused output). All the output cables from the splitter are the same length. The synchronization pulse repeats with a known period greater than one second which is set from software on start up.

Each SNAP2 board must undergo a synchronization procedure to enable time-stamping its data. On startup, the first SNAP2 board receives a 64-bit integer indicating the UTC time that the next PPS corresponds to, expressed in number of clock cycles since the Unix epoch.

When each of the other SNAP2 boards go through their synchronization procedure (which need not be done simultaneously for all the boards), they receive the 64-bit integer timestamp that will correspond to the next pulse from SNAP2 #1. When the next pulse from SNAP2 #1 arrives, it triggers a synchronization pulse that propagates through the FPGA logic to all the firmware components that require a time reference. Within each of these subsystems, on receipt of the synchronization pulse, a counter begins counting up from the received Unix time value, incrementing by one on each clock cycle. The output of these counters can be used to timestamp the data in that firmware subsystem (see section 3.6 for the application to the cosmic ray firmware).

The values of these 64-bit integer timestamps from all the SNAP2 boards thus refer to one time reference, known as the Telescope Time, from which the relative time of events at different antennas can be estimated. Absolute timing then requires accounting for any offsets between the Telescope Time and absolute external references. For the purposes of the cosmic ray detector, the synchronized Telescope Time is a sufficient clock since cosmic rays will not need to be pinpointed to source locations on the sky (which would require tying to an absolute global time reference).

2.5 Communication between SNAP2 boards for whole-array snapshots

Overview

For every trigger that is not vetoed, the cosmic ray detection system saves snapshots of data from the entire array, including antennas for which the pulse was below the detection threshold. This section details the communication between the SNAP2

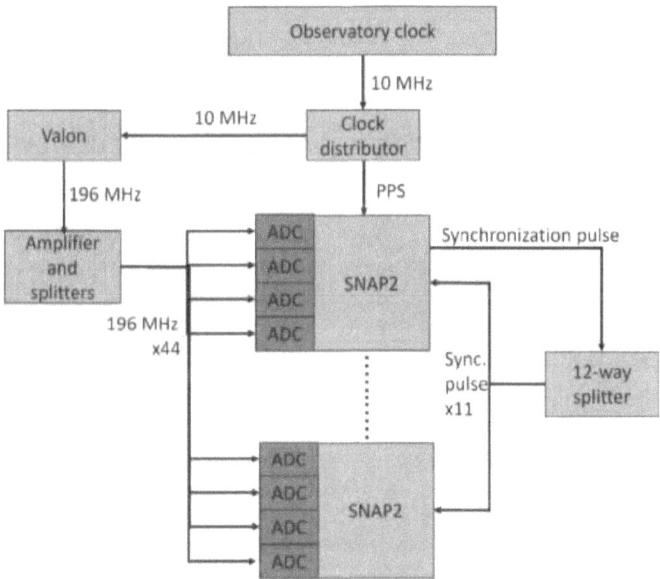

Figure 2.7: Clock distribution system for the OVRO-LWA. A 10 MHz signal from the main observatory clock also transmits a 1 Hz signal by missing one pulse every second. The clock distributor outputs a PPS signal and a 10 MHz reference separately. The Valon signal generator outputs a 196 MHz sinusoid which is split to all of the 44 ADCs cards using 44 cables of the same length. The PPS signal is sent to one SNAP2 board, which distributes a synchronization pulse to all the others and back to itself via a splitter, again with all 11 cables the same length. Adapted from a diagram by Jack Hickish, with permission.

boards that accomplishes data readout from the whole array. Since each SNAP2 processes data from a set of 32 antennas, trigger conditions are identified based on coincidences between subsets of 32 antennas, and the signal to trigger data readout must be communicated between from the SNAP2 that originates the trigger to all the other SNAP2s.

Ethernet is asynchronous. The devices on the network operate at different clock rates without a common reference. If an FPGA sends a packet on a certain clock cycle, it is not certain how many FPGA clock cycles later the packet will emerge from the network switch. For communication between boards that is faster than Ethernet and has a fixed, predictable latency, I use the SNAP2 boards' general-purpose input

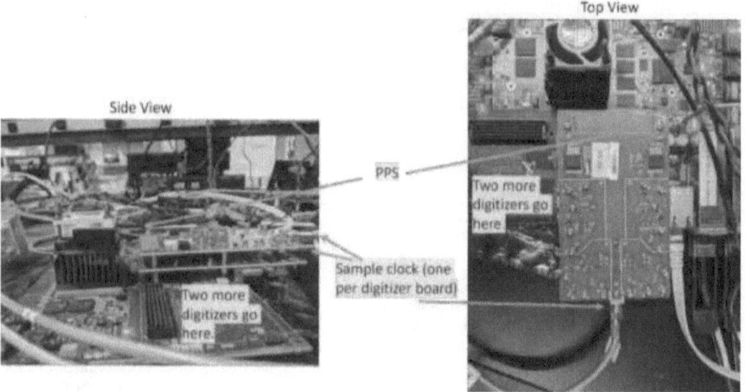

Figure 2.8: SNAP2 board with clock inputs highlighted. These images are from the lab bench test setup because the SNAP2 boards in the digital electronics rack at the OVRO LWA are packed to close to easily view the highlighted features. On the right, the two SMA inputs (with yellow covers) below the PPS are used to distribute the PPS to the other SNAP2 boards (one sends the PPS and the other receives).

output (GPIO) pins. The SMA inputs are not available for this purpose as they are used by the clock distribution system. The GPIO pins are located on a 10-pin header, two of which are available for general-purpose input output signalling from the FPGA. I use one of these two pins as an input pin, for receiving trigger signals and the other as an output pin for transmitting trigger signals. Since there aren't enough pins for every SNAP2 to have a dedicated connection with every other SNAP2, the boards are linked in a loop with wires that carry a one-bit trigger signal such that each board's trigger output pin sends a signal to a neighboring board's trigger input pin. When any board originates a trigger from its own search logic or receives a trigger (either from its neighbor or from software), if it is not already in readout state then it goes into readout state and passes the trigger to the next board. If it is already in readout state, then it ignores the trigger, in order to prevent the trigger signal being passed around the loop indefinitely.

Initial design and reasons for re-designing

In the OVRO-LWA electronics shelter, the 11 SNAP2 boards are mounted edge-wise in two chasses, which are stacked one above the other in a computer rack. Six SNAP2 boards sit in one rack and five in the other (See Figure 2.9). I began working on the communication between the SNAP2s in a test setup on a lab bench, before the OVRO-LWA electronics shelter was ready for the SNAP2 boards to be installed. In the bench setup, the SNAP2 boards were lying flat on the bench and did not have ADC cards installed, making access to the GPIO pin header easier. Among the other pins in the header, one is ground, one is a fixed 3.3V, two have a voltage that tracks two of the LEDs on the board and could be controlled by the firmware, two are GPIO pins tied to the Microblaze CPU within the Ultrascale+ FPGA, and two are GPIO pins connected to the (smaller) Xilinx Zynq FPGA. I initially used insulation displacement connectors to connect wires to the GPIO pins. I also initially programmed the FPGAs to use a very short pulse as a trigger signal on the wire. It was necessary to extend the pulse so that the voltage stays high for many clock cycles in order to transmit the signal. I currently use 1000 clock cycles.

The main lab test I used involved connecting three SNAP2 boards in a loop, sending a software trigger to one board and then measuring (via a counter in the firmware) the clock cycles until the trigger makes it back to that board. In these tests, no ADCs are connected (and the event search firmware was not running) and the only possible sources of a trigger signal are a signal from software or from another SNAP2. The steps to execute the test were as follows:

1. Link GPIO pins for a loop of three boards.

2. Reset the trigger counter on the FPGA.

3. Reset the trigger timer.

4. Set the board state to listening.

5. Read the readout state register.

6. Send a trigger to one board from software.

7. Read the register that counts how many clock cycles between subsequent triggers. This is only meaningful for the first board, since it will count the initial trigger from software as well as a second count when the trigger signal returns around the loop. The other boards should only count one trigger.

8. Read the register that counts the number of triggers the board received. This should be two for the first boards and one for the other boards.

9. Verify that the readout state behaved as expected.

10. Reset the counters.

11. Repeat 500 times.

Counting the number of triggers each board saw checks for ringing. Oscillations on the wire could cause the voltage at the In pin to cross the threshold multiple times and be logged as multiple triggers. Figure 2.10 shows the results of these tests for one of the first connectors tested. If there were no ringing, only board 9 would count two triggers, because it received the initial trigger, and the other boards would see one trigger. The other boards occasionally count extra triggers, suggesting ringing. The small amount of ringing seen here (board 11 usually saw two triggers and board 12 sometimes did) may have been acceptable, since the FPGA logic ignores repeated triggers for the amount of time it takes to transmit the buffered data. However, with longer wires (as would be needed in the electronics shelter) the ringing became a more serious problem, and oscilloscope traces showed very large oscillations.

Furthermore, the connector pictured above was not very strong, made it difficult to remove a SNAP2 from the rack, and was very difficult to fit in the tight clearance under an ADC card. Hirose connectors had better clearance and made it easier to disconnect and re-connect SNAP2 boards, but had worse trouble with ringing and a less reliable electrical connection.

Larry D'Addario identified that a bidirectional amplifier chip between the GPIO pin header and the main FPGA couldn't drive a clean signal down a long wire because the impedance of the wire causes the bidirectional amplifier to oscillate, and designed an auxiliary PCB to transmit the signal from the GPIO pin down a long wire. Figure 2.11 shows the auxiliary circuit as well as a 3D-printed mock up to test the clearance under the ADCs. Each signal wire is twisted with a ground wire for shielding, and the ground wires connect to the ground pin on the pin header.

The auxiliary circuit successfully transmits triggers with no ringing behavior, as measured by counting triggers received in firmware and inspecting oscilloscope traces (See Figure 2.12). Additionally, I ran the trigger counting firmware overnight to confirm that there were no spurious triggers, and I repeated tests for wires up to a meter long. I repeated the trigger-timing and counting test when all the SNAP2 boards were installed at the OVRO-LWA. In the final system, the trigger takes 137 clock cycles (occasionally arriving close to the clock cycle boundary such that it is effectively 136 cycles) around the entire loop of 11 SNAP2 boards.

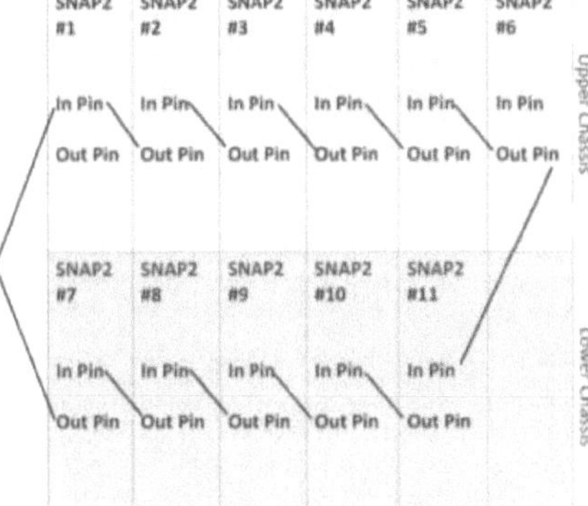

Figure 2.9: Left: Diagram of the SNAP2 board chassis, showing the interconnection between SNAP2 boards. Right: Photo of two SNAP2 boards installed in the OVRO-LWA digital signal processing electronics rack. The photo is taken with only two boards installed to make it easier to see. Only the SNAP2 on the far right has all of the input signals connected to its ADCs. The communication wire between the SNAP2s is the green and white twisted pair in the lower left.

2.6 Analog Signal Path for Outrigger Antennas

Signal attenuation makes coaxial cables impractical for bringing RF signals more than a couple hundred meters from the antenna. Inserting stages of amplifiers along the signal path would amplify the noise as well, and would not address the dispersion of the signal due to the frequency-dependent index of refraction of the cable. One option for large arrays is to place the analog-to-digital converters near each antenna and transmit the digitized signal back to a central correlator over optical fiber. In the extreme case of very long baseline interferometry, raw data recorded at each antenna is simply shipped on hard disks to a central location for correlation.

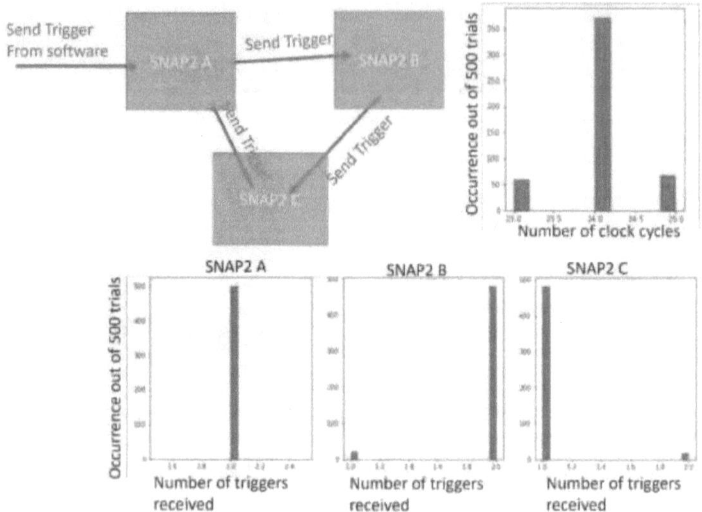

Figure 2.10: Test of communication stability with the first connectors tested. Upper left: Diagram of the test setup using three SNAP2 boards. A trigger is sent from software to SNAP2 A, which passes it to B, which passes it to C, which passes it back to A. Upper right: Distribution of the number of clock cycles for SNAP2 A to receive the trigger signal back from C. Lower three plots: Distribution of the number of triggers received by each SNAP2. Only SNAP2 A should count two triggers since it received the initial trigger from software. Any additional triggers (beyond one) counted by the other SNAP2s indicate ringing in the trigger signal.

Another option is a technique that transmits analog radio-frequency signals over optical fiber (RFoF) to a central location where the signals are digitized, and this is the technique used by the OVRO-LWA. The RF signal from each dipole is input to a circuit by which it modulates the current to a laser diode, such that the intensity of the laser light modulates at radio frequency. When coupled to an optical fiber, the RF-modulated laser light can be transmitted long distances. At the other end of the optical fiber, a photodiode receiving the light outputs an RF signal.

The advantages of RF over fiber include moving more of the signal processing electronics away from the antennas to reduce the risk of generating RFI near the antennas, and allowing digital signal processing to occur in a central location. Keeping digital signal processing in a central location avoids having to weather-proof and supply power to digitizers near each antenna, and also simplifies the

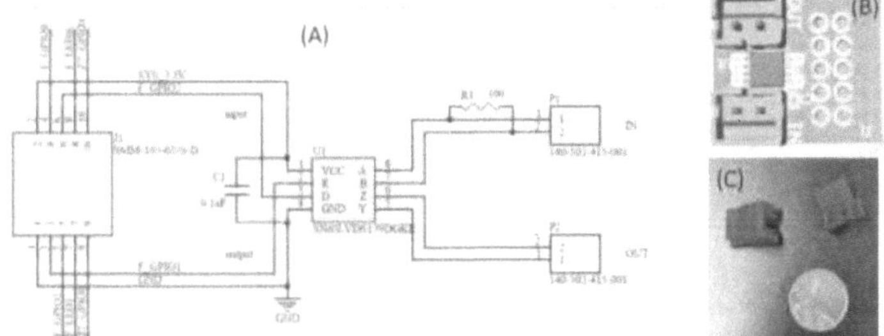

Figure 2.11: Auxiliary circuit used to transmit a reliable trigger signal down longer wires. (A) Circuit designed by Larry D'Addario. (B) CAD image of the circuit. Image from Larry D'Addario. (C) 3D printed mock-ups of the auxiliary board used to test clearance under the ADC cards. The two are printed from opposite orientations so the defects from support plastic are on opposite sides.

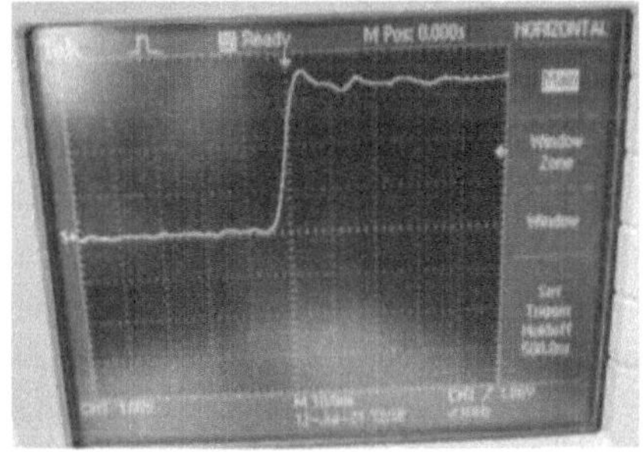

Figure 2.12: Left: Oscilloscope trace of the trigger signal using the auxiliary circuit, showing negligible ringing that rapidly decays to a stable voltage. The horizontal lines mark 1-Volt increments. Left: Photograph of the measurement setup.

distribution of clock signals to each digitizer. Disadvantages include challenges to maintain gain stability and linearity, particularly in the presence of strong RFI. RF over fiber is far from ubiquitous but is used by several recently-built radio astronomy arrays (e.g. HERA [83], UTMOST2D (Mandlik et al. 2023 in prep), DSA-110 [82]) and is an option under planning for the upcoming Square Kilometer Array [43].

The use of RF over fiber at the OVRO-LWA is crucial to the chosen cosmic ray detection strategy. Since the initial stage of cosmic ray detection takes place on the FPGAs that process the raw output of the analog-to-digital converters, bringing all the signals to a central location enables the strategy of using distant antennas to veto RFI. The use of RF over fiber allows FPGAs that handle signals from core antennas to also include a few of the most distant antennas, to veto events which illuminate the entire array.

The fiber optic laser driver modules were redesigned by James Lamb for the upgrade to the OVRO-LWA. Drawbacks to the previous modules included a difficult-to-calibrate temperature-dependent gain and phase response. Since the Owens Valley is a high desert, temperature can vary by several tens of degrees in a day. As part of the upgrade, laser diodes for the new laser driver modules were individually tested to characterize their temperature-dependent response. This temperature-dependence varies between laser diodes, even between diodes that were likely cut from the same semiconductor wafer. For the upgraded array, a large number of extra laser diodes were tested so that the best subset with similar temperature response were selected to be used in the array.

Figure 2.13 shows an example of a test setup for investigating the temperature dependence of laser driver modules. In this example, we measured optical optical power and S21 for stage II laser driver modules as a function of temperature. We used a Cincinnati Sub Zero fridge-oven to test the laser driver from -30 C to 40 C in steps of 10 C. In the test shown in Figure 2.13, the RF input of one channel is driven by port 1 of a Rohde and Schwarz vector network analyzer (VNA), which sweeps a tone from 0 to 25 GHz. The fiber output of that channel is connected to port 2 of the VNA via a diode and bias T. With this setup, the VNA can measure the S21 parameter as a function of frequency. Before beginning, we did a Transmission, Open, Short, Matched (TOSM) calibration with the calibration set placed at the end of the SMA from port 1, and at the input to the bias T at port 2. Thus, the port 1 measurement is taken at the input to the laser driver and the port 2 measurement is taken at the input to the bias T. The optical fiber connectors were examined

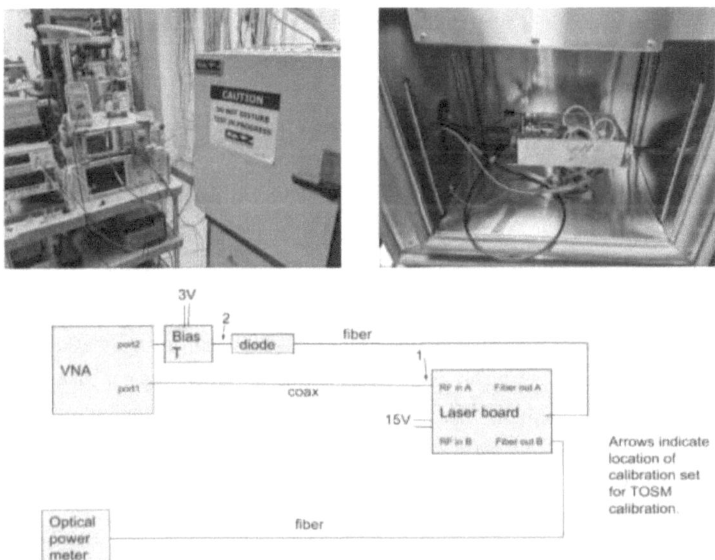

Figure 2.13: Fiber optic laser driver module testing. Upper left: Laboratory test set up. The laser driver module is inside the temperature-controlled chamber (in the right of the photo). One channel of the laser driver module connects to the VNA (center of photo) via a photodiode and a bias T, and the other connects to an optical power meter (left in photo). Upper right: Photo of a laser driver module in the temperature-controlled chamber with the chamber door open. Signal cables, fiber, and power wires pass through a port in the left side of the chamber. The laser driver module pictured is from the Stage 2 array, and is tested here as part of the upgrade. Lower: Diagram of the test setup.

with a ViewConn device and cleaned before connecting. The test shown here was conducted in April of 2022.

2.7 Simulating the Impact of Antenna Placement near the Core Array for Cosmic Ray Detection

Motivation

The choice of antenna layout for the OVRO-LWA was primarily driven by requirements for interferometric imaging, but cosmic ray detection considerations contributed to the decision to place additional antennas around the outside of the dense core array. This section describes a set of simple simulations for comparing several proposed array configurations while the array layout was being finalized in early 2020.

Detecting a cosmic ray requires that the air shower illuminate more than the threshold number of antennas. A highly confident detection, however, does not guarantee a good reconstruction, particularly if the detecting antennas are spatially clustered in a small region of the air shower footprint. Typically, the more thoroughly an air shower footprint is spatially sampled (i.e. detected at antennas in different regions of the footprint), the more precisely the observables can be reconstructed.

Computationally-intensive ZHaireS simulations of numerous air showers can precisely estimate the quality of air shower reconstructions for a proposed array, but to make quick comparisons of antenna configurations when the array layout was being finalized, I needed a simple approach that computes a less-computationally-intensive figure-of-merit given a set of proposed antenna locations. In this section, I describe a simple geometric simulation and present the results of a comparison of four array configurations. I did this simulation work in March of 2020, when the lab was closed due to the pandemic.

Simulation

I model air showers as cones of light that illuminate antennas within the footprint described by the interior of the ellipse at the intersection of the cone and the ground. I consider an antenna to have detected a cosmic ray if the antenna lies within that cosmic ray's footprint. For each air shower cone, the number of core antennas in the footprint determines whether that air shower is detectable, and the number of antennas within the footprint but outside the dense core array is used as a proxy for the quality of the spatial sampling of the footprint. The simulated cosmic rays are uniformly distributed in azimuth and core position, and distributed in Zenith Angle according to the distribution observed by Monroe et al. [65].

Optimizing for Lower Energy Cosmic Rays   The optimal array configuration
for observing 5×10^{16}eV cosmic rays (the lower limit of our energy band) is different
from the optimal array configuration for observing 10^{18}eV cosmic rays (the upper
limit of we expect to detect). My simulation produces one figure-of-merit optimized
for cosmic rays in the mid to low end of our energy band. I make this choice because
the sample of 10^{18}eV cosmic rays is less important to our goal of comparing events
across the Galactic to extra-galactic transition, is a small fraction of our expected
total sample, and is less sensitive to the types of changes in antenna placement
currently under consideration.

Specifically, I have made two choices that optimize the simulation for lower-energy
($\sim 5 \times 10^{16}$–5×10^{17}eV) cosmic rays and exclude higher-energy ($\sim 10^{18}$eV) cosmic
rays:

1. Requiring that the core array detects the cosmic ray:
My simulation considers an air shower to be detected by the array if at least six core
antennas detect it. A six-antenna detection is the threshold for the trigger firmware.
In principle, a detection by six outlier antennas would suffice. However, only highly
inclined air showers have a large enough cross-section to illuminate this many outlier
antennas, and only very high-energy cosmic rays make a bright enough air shower to
be detected at large inclinations. Thus, air showers for the energies I am prioritizing
must land on core antennas in order to trigger the array.

2. Limiting the Zenith Angle distribution:
To generate simulated air showers, I use the distribution of zenith angles observed
by [65], but restricted to the < 35 degree tail for my simulations. As described
above, only very high-energy cosmic rays can be detected at larger zenith angles.
The 35 degree cutoff ensures that the modeled air showers are suitable down to the
lower limit of our energy band.

Proxy for Reconstruction Precision   Even one antenna on an otherwise-
unsampled side of a cosmic ray footprint can greatly improve the quality of the
parameter reconstruction. Unless an air shower arrives from zenith over the center
of the core (which is an excellent, but rare scenario), then the core antennas sample
only one side of the shower. Illuminating even a just a few outlier antennas near the
core improves the reconstruction. My simulation uses the number of outlier antennas in the footprint of a detected air shower as a proxy for how well that air shower
is reconstructed. The simulation generates 10000 cosmic ray air shower cones and

outputs the number of outlier antennas withing the footprint of each cosmic ray that is successfully detected by the core array. The cumulative distribution of this output can be used to compare different array configurations.

Geometric Model   Generating a footprint ellipse for an air shower requires the cone opening angle, the distance to the apex of the cone, the center position, the azimuth of the semimajor axis, and the inclination (zenith angle) of the air shower. The main simplifying assumptions I make in order to determine the geometry of the generated air showers are:

1. All the air showers are cones whose apex corresponds to $X_{max} = 680 \text{g/cm}^2$, typical of a 10^{17} eV proton. (X_{max} is the atmospheric column density at the peak of the particle interactions in the air shower).

2. All the air showers have the same opening angle (2.27 degrees). I determined this angle from comparison to a ZHaireS simulation of the air shower footprint of a 10^{17} eV proton with $X_{max} = 680 g/cm^2$ (Washington Carvalho, private communication). In the ZHaireS-simulated footprint, all antennas within a ~ 100 m radius of the shower center experienced an electric pulse above their sensitivity limit. To obtain an opening angle for the cone approximation, I calculated the vertical distance to $X_{max} = 680 g/cm^2$ by assuming a slab atmosphere with an exponential density profile. The height and width determine the opening angle.

The azimuth is drawn from a uniform distribution (a more detailed simulation could include the small effect of the Earth's magnetic field). The center position is drawn from a uniform distribution in a 1 square kilometer box centered on the array core. The zenith angle is drawn by sampling the observed Gaussian distribution, with a cutoff imposed at 35 degrees (described above).

The zenith angle determines not only the eccentricity of the ellipse but also the distance to the cone apex. A more inclined cosmic ray reaches a given atmospheric column density higher in the atmosphere than does a cosmic ray arriving from zenith. I assumed a slab atmosphere with an exponential density profile and derived the position of the cone apex as a function of zenith angle, for fixed X_{max}.

The distance, L, to the apex is:

$$L = \sec(ZA) h_s \log\left(\sec(ZA)\frac{X_{tot}}{X_{max}} - h_{\text{OVRO}}\right)$$

where ZA is the zenith angle, $h_s = 9020\,\text{m}$ is the pressure scale height of the atmosphere, $X_{tot=1030g/cm^2}$ is the total vertical column density, $X_{max} = 680g/cm^2$, and $h_{OVRO} = 1219\,\text{m}$ is the elevation above sea level of the observatory.

Figure 1 shows an example simulated footprint over the original planned LWA-352 configuration. The gray shaded region is the elliptical footprint of an air shower with a zenith angle of 32 degrees, landing South-West of the array core. Orange points are outlier antennas, blue points are core antennas, and red points mark antennas that would detect this cosmic ray. In my simulation, this air shower would be detected because 34 core antennas lie in the ellipse (well above the 6-antenna threshold), and it would be added to the output with a value of 3, since three outlier antennas lie in the ellipse. This footprint is an example of a cosmic ray whose key observables could likely be reconstructed well.

Comparison of Four Arrays

I use my simulation to compare four array configurations, shown in Figure 2. The first configuration is the original LWA-352 plan. The empty area to the south of the core means that many cosmic ray footprints illuminating the south side of the core array would not illuminate any outlier antennas, and for these events the reconstruction of key observables would be poor or impossible. The second configuration adds three outlier antennas to the ring around the core, filling in the gap in the South. The third configuration is the configuration David Woody made in March 2020. The fourth configuration adds just one antenna to this new configuration, filling in a gap near the core to the south. Figure 3 compares simulation results for these three configurations.

The right panel of Figure 3 is the main result. It shows the number of cosmic rays detected by N or more outlier antennas as a function of N, for each of the four array configurations. For both the original and new configuration, filling in gaps in the South makes a significant improvement. However, the improvement from the original proposed configuration to the new configuration is even larger. The best of the four configurations tested is the recent configuration with one Southern antenna added. This configuration more than doubles the number of simulated cosmic rays air shower cones detected with three or more outlier antennas, compared to the original configuration.

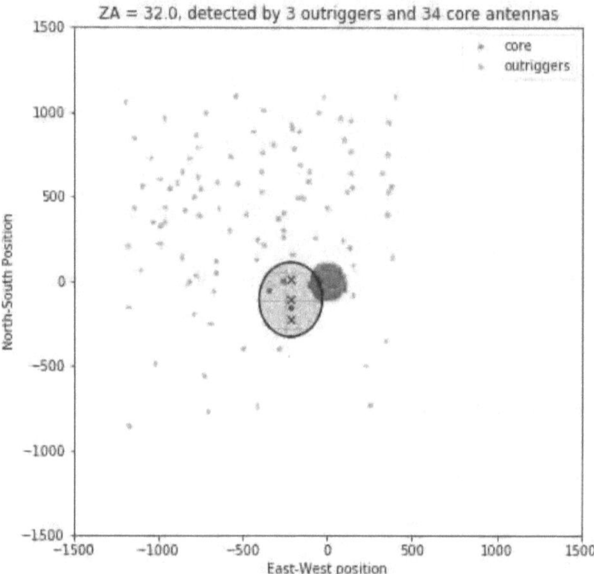

Figure 2.14: Simulated footprint for an air shower with a zenith angle of 32 degrees. Orange points mark the positions of outlier antennas, blue points mark core antennas, and red point mark any antenna that could detect the cosmic ray. The gray shaded region is simulated footprint. Black crosses mark the center and foci of the ellipse. The x and y axis limits extend to the boundaries of the box used in this simulation.

Conclusions

Filling in gaps in outliers in the South improves the number cosmic rays for which a good reconstruction is possible. In general, outlier antennas near the core are more important than distant outliers. This is the qualitative reason that the new configuration is such a significant improvement for cosmic ray science, compared to the original LWA-352 plan. This simple geometric simulation is fast to run (a few seconds for 10000 cosmic rays), making a convenient tool for quick comparison of array configurations. It neglects the spatial variation of intensity within the air shower footprint, and more detailed simulations could be run with ZHaireS, but the geometric simulation captures key factors for comparing configurations, with a

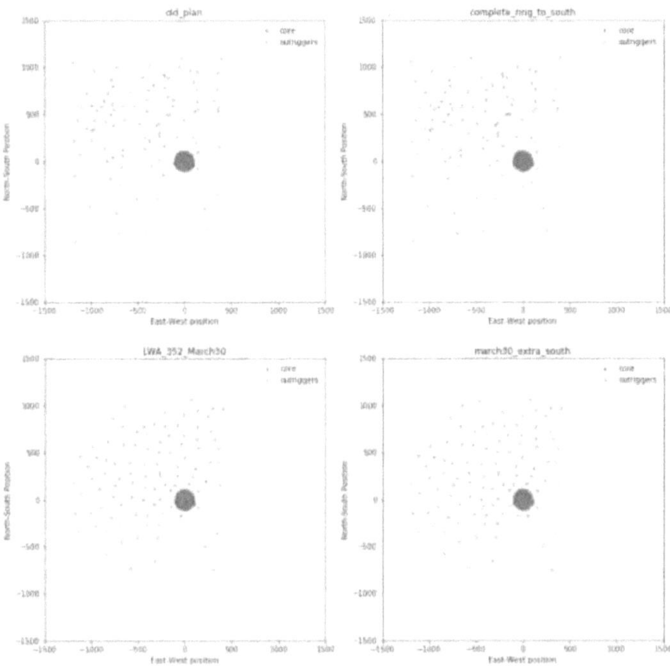

Figure 2.15: Four array configurations for comparison. The upper left is the original plan. The upper right adds to the original plan three antennas in the ring around the core, filling in the gap in the South. The lower left is the new configuration made by David Woody in March 2020. The lower right adds one antenna to this new configuration, filling in a hole in the South. In all the plots, the orange points are outlier antennas and the blue plots are core antennas.

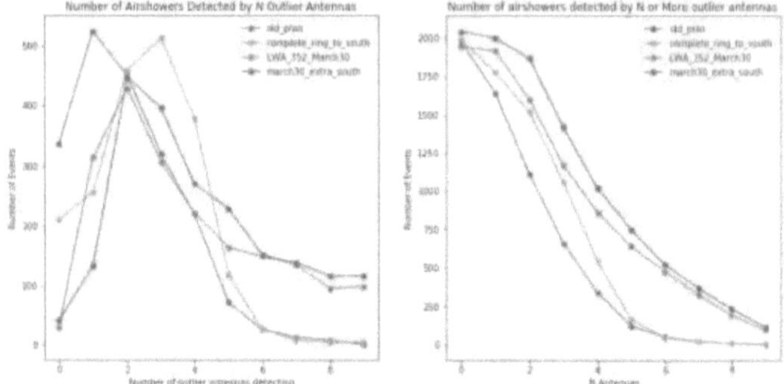

Figure 2.16: Simulation results for the four array configurations shown in Figure 3. The left panel is a histogram of the number of detectable air showers sampled by different numbers of outlier antennas. The right panel shows the cumulative distributions integrated under the upper tails of the distributions in the left panel. In other words, it shows the number of cosmic rays detected by N or more outlier antennas, for different values of N.

particular focus on the energy range of interest. Informed in part by these results, the final array layout (see Figure 2.2) partially fills in the open area to the south of the core array, compared to the original proposed layout.

2.8 A brief investigation on the importance of keeping the antennas free of vegetation

This section describes a brief test to investigate whether vegetation around the antennas affects measurements with the antennas. In November of 2021 several of the core antennas (at that time only core array antennas had completed the upgrade) that had intermittent unusual signal properties (including intense spikes in amplitude) had been noticed to have vegetation growing through the ground screens and touching the antennas. Much of the vegetation in the LWA is sagebrush and tumbleweed, and these particular antennas had a different plant which was likely Mojave indigo bush (Psorothamnus *arborescens*), and was noteably less dry (and thus likely dielectrically different from the others). I wanted to investigate whether green plants touching the antennas were responsible for unusual signal properties. By the time of the test, the vegetation originally noticed had been removed. I found two antennas with similar vegetation: antennas number 45 and 9. On antenna 45, the plants mainly touched the N-S dipole. To test the affect of the vegetation, I recorded 500 snapshots triggered from software readout at a rate of 2 Hz. Every set of 500 snapshots alternated between having the vegetation pushed against the N-S dipole of antenna 45 vs held away from the dipole.

Figure 2.18 shows images of antenna 45. Note the black marks where plants touched the metal. This was present on other antennas with plants too.

The following plots overlay the results of all trials, by showing the standard deviation and excess kurtosis of each of the 500 snapshot time series in the trial. Color identifies the dipole. All the trials with the vegetation touching the North-South dipole of antenna 45 are solid lines and all the trials with the vegetation pulled away are dotted lines.

Antenna 9 had data recorded and is included in the plots because it also had a large shrub of likely P *arborescens*, but the tests did not control when that vegetation did or didn't make contact with the antenna.

Differences in the kurtosis plots are more notable than standard deviation. All the largest kurtosis spikes occur in trials with the plant touching, and for the dipole the plant was touching. However, only the first two trials with the plant touching have these large spikes—the last trial does not.

The vegetation may potentially cause a difference in the sensitivity to RFI by putting the antenna in electrical contact with the ground screen and changing the beam pattern. It is not clear why the spikes in excess kurtosis did not appear in the last test. Potentially the ambient RFI events were different during that time. Although a larger number of trials and tests with more antennas would be certainly be useful to confirm this affect, these preliminary results contributed to the decision to keep vegetation away from the antennas to the extent possible.

2.9 Conclusion

The upgrade to the OVRO-LWA took place over four years, from a preliminary design review in 2019 to the completion of the final antenna in 2023. The layout of the array, with a dense core and distant outliers enables a cosmic ray detection

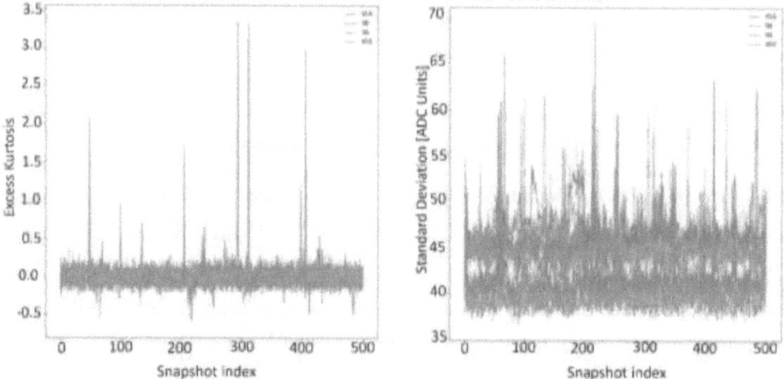

Figure 2.17: Left: Excess Kurtosis of each timeseries snapshot plotted in chrono-
logical order as a function of snapshot index. Color indicates which of four dipoles
the signal is from. Data from all six trials are overlaid. Solid lines indicate data
from trials with the vegetation touching dipole 45A. Dashed lines indicate data from
trials with the vegetation pulled away from 45A. Right: Same as left but showing
standard deviation instead of kurtosis of each snapshot.

Figure 2.18: Left: Image of antenna 45 with shrub growing in it. Center: Close up
photo of dipole 45A. Note the marks on the antenna. Right: Antenna 9.

strategy that searches for air showers illuminating the core array and uses the most distant antennas to veto RFI. This approach to a radio-only trigger is possible because the signals from all antennas are digitized in a central location. RF-modulated laser light brings signals from the most distant antennas to this central signal processing location on optical fiber. Having large groups of signals (both polarizations of 32 antennas) processed within a single FPGA supports a radio-only cosmic ray search strategy of searching for events that coincide among neighboring core antennas and are not detected by the veto antennas. Good synchronization between the FPGA boards is expected to allow X_{max} reconstruction with an uncertainty of $17\,\mathrm{g/cm^2}$, and a fast, reliable communication link between the FPGA boards allows data to be recorded from the entire array when one board initiates a trigger signal. The next chapter will go into detail about the design and implementation of the cosmic ray trigger firmware.

Chapter 3

FPGA FIRMWARE DESIGN FOR A RADIO-ONLY COSMIC RAY TRIGGER AT THE OVRO-LWA

This chapter describes the cosmic ray firmware in detail, with two goals: (1) to describe the firmware at a level detail suitable to facilitate future work building on this design, and (2) to discuss the reasons behind design decisions.

The cosmic ray search runs on Field-Programmable Gate Array boards (see Chapter 2) that process the raw output of the ADCs that digitize the radio signals. The FPGA firmware implements an event search state and a data readout state. It buffers 20 microseconds of data while searching for events that illuminate core antennas and blocking events that illuminate distant veto antennas. When any boards' antennas meet the trigger conditions all read out their data. The 20 microsecond buffer ensures an ample record of background conditions before the event and also ensures that regardless of the event arrival direction, a window of data corresponding to the light travel time between the antennas does not approach the edge of the buffer.

Section 3.2 gives a high level overview of the main parts of the OVRO-LWA cosmic ray firmware and section 3.3 describes the inputs and outputs of the system. The three main subsystems are described in sections 3.4 (state control), 3.5 (trigger and veto), and 3.6 (buffer and data readout). Section 3.7 overviews the software control of the cosmic ray firmware, section 3.8 discusses tests to characterize the data rate that the system can handle, and section 3.9 provides a concluding discussion of design considerations. In order for this book to thoroughly document the firmware design, I include figures of simplified logic flow-chart diagrams as well as screenshots of the corresponding portion of the Simulink diagram, so that this chapter can be used as a manual for understanding the Simulink file in detail.

3.1 Introduction to synchronous digital signal processing with FPGAs

Field-programmable gate arrays (FPGAs) function as reconfigurable logic circuits. An FPGA consists of a network of hardware components (hence the "gate array" part of the name) that perform various logic or arithmetic operations, or store data as random access memory (RAM). Much like an application-specific integrated circuit (ASIC), these components can be connected to each other to build complex computations, and connect to input and output pins that interface with other hardware devices (such as analog-to-digital converters (ADCs) and Ethernet ports). Unlike an ASIC, the components of an FPGA can be reconnected in different ways (hence the "field programmable" part of the name). The coded set of instructions for configuring an FPGA is often referred to as firmware because it is neither hardware design nor software code.

Advantages of FPGAs include handling extremely large input and output data rates, and the ability to be configured to have fast, low-level interfaces with other hardware devices, such as ADCS. Additional important characteristics include synchronous computations parallel processing. FPGAs perform computations synchronously—every gate in an FPGA has an input and an output value on every clock cycle, outputs of one subsystem connect to inputs of others, and so the firmware specifies what happens on each clock cycle, and every component can perform its operation on every clock cycle. Thus, FPGAs can accomplish many types of computations much faster than CPUs, even though typical clock rates are much lower for FPGAs. The Xilinx Kintex Ultrascale+ FPGAs in the SNAP2 boards in the OVRO-LWA receive a clock signal synchronized to the ADC sample rate of 196 MHz. This clock rate is much slower than the GHz clock rates typical of modern CPUs, but a CPU takes more clock cycles to perform each operation, and takes additional time to maintain the capability of switching between multiple tasks. Since every gate operates on every clock cycle, FPGAs can readily be configured to perform operations on multiple data streams in parallel, simply by configuring parallel sets of logic. The large number of input and output pins can be configured to receive or drive signals that connect to other hardware components. All these features of FPGAs make them excellent for interfacing with analog to digital converters, and so FPGAs perform the first stages of digital signal processing in most modern radio telescopes.

Programming an FPGA is different from software programming. Whereas a software compiler takes a description of what the program should do and attempts to find the best way to implement that on the hardware architecture available, programming an FPGA requires finding the right arrangement of the available hardware resources to build a circuit that acts according to the description. Synchronous computations necessitate consideration of the timing of every computation, which results in component placement constraints. If the output of one component is input to another component, then those components must be placed near enough to each other that the signal can travel from one to the other well within a clock cycle. Every operation has some latency, and many operations consist of a group of lower-level operations, each with their own latency.

For blocks consisting of many logic gates and sub-components, it is typically important to know how many clock cycles go by before the output corresponds to a given input. Keeping different components synchronized requires specifying additional delays on certain signals as needed. In this chapter, when I refer to a signal or a timeseries or a stream of data, I mean the digital numbers at a particular place in the logic design, that takes one new value each clock cycle. The timeseries of ADC voltage data for a particular antenna are the series of numbers output by the ADC channel corresponding to that antenna. The filtered power timeseries computed from those voltages are another series of numbers appearing one clock cycle at a time at a different place in the FPGA, with some time offset (due to the latency of the computation) compared to the ADC voltage data from which they are computed.

Firmware for the OVRO-LWA is compiled using the CASPER toolflow (e.g. [49]) which enables firmware designs to be described graphically in a readily human-readable way and then compiled to a bitstream that can be loaded onto an FPGA to configure its logic fabric to function as the circuit that the high-level diagram represents. This toolflow uses Matlab Simulink to describe the firmware logic flow graphically in a series of interconnected "block" components. Each block has inputs and outputs that have a value on each clock cycle, and the block contains a description of operations that will be performed to relate the inputs to the outputs. Blocks are often described hierarchically, such that blocks consists of other blocks, which ultimately fundamentally consist of wrappers around code in Very High-Speed Integrated Circuit Hardware Description Language (VHDL) that can be transformed into a hardware configuration. These lowest-level blocks are drawn from either the Xilinx library or the CASPER XPS library. The Xilinx library contains blocks for operations such as counters, arithmetic, logical comparisons, delays, addressable RAM, etc. Blocks from the CASPER DSP library consist of Xilinx blocks combined in commonly-used ways (such as to build fast Fourier transforms). In contrast, the CASPER XPS library contains blocks that are wrappers around VHDL descriptions involving specific pins on the circuit board for which the design is intended—such as interfaces with Ethernet hardware, software registers, and general-purpose input and output pins.

The CASPER toolflow compiles the Simulink diagrams to a VHDL description of the design, and then uses Vivado to create a bitstream which can be used to program the FPGA to operate as the circuit described in the design. The Vivado software handles placing the components in a map of the FPGA fabric to attempt to optimize resource use and to keep the latency within the constraint set by the intended clock rate, i.e. if the output of one block is input to another block, the route connecting those blocks must be short enough for the signal to arrive before the next clock cycle.

Making CASPER firmware for specific hardware (such as the SNAP2 boards) typically requires very specific versions of the associated software tools. I compile firmware using Vivado version 2019.1, Matlab version R2019a, and Jack Hickish's Caltech-LWA version of the CASPER library[1] (see Appendix A for details on setting up the software environment).

3.2 Overview of the cosmic ray firmware

The cosmic ray firmware has three main subsystems (see Figures 3.1 and 3.2): (1) a state control subsystem that manages whether the SNAP2 board is in a readout state or is searching for new triggers, (2) the trigger and veto logic, and (3) the buffer and Ethernet packetizer. Other parts of the firmware include internally-generated test signals, and logic for software registers and block random access memory (BRAMs) to control and monitor the system's behavior. The three main subsystems are detailed in individual sections of this chapter:

- State control, part D of Figure 3.1 and section 3.4

- Trigger and veto, part G of Figure 3.1 and section 3.5

- Buffer and packetizer, part E of Figure 3.1 and section 3.6

Other elements of the top-level cosmic ray firmware design

A few components, which I document here, sit outside the three main subsystem blocks. I keep all the blocks that interface with input and output hardware at the highest level of the cosmic ray Simulink diagram (this choice is based on a CASPER convention to put such blocks at the top level of a Simulink diagram, although in this case they are within the cosmic ray subsystem rather than the top level of the entire design). These blocks are all indicated with yellow in Simulink, and include blocks such as software registers (memory that is readable and/or writable from software) and Ethernet blocks. Parts A and B in Figure 3.1 are the software registers and general purpose input output pin interfaces. Software registers are 32 bits, but many of the quantities to be read or written to and from these registers do not need so many bits. Thus, to reduce IO usage, several smaller integers can share a software register, and slice blocks separate the output of the software register into its separate components.

Another yellow block in the top-level cosmic ray Simulink diagram is the 40 Gb Ethernet block (F in Figure 3.1). Conceptually this block is part of the buffer and packetizer subsystem, but due to the convention of placing yellow blocks at the top level sits outside of the subsystem block (E in Figure 3.1) which accordingly outputs the required inputs to the Ethernet block. Similarly, a number of event-rate-logging Shared Block RAMs sit outside the trigger and veto subsystem block. The subsystem block is the white block in the left of region (G) in Figure 3.1), and the event-rate-logging Shared BRAMs are to the right within (G).

Parts (C), (H), and (I) in Figure 3.1 are at this level of the diagram because they are not specifically associated with one of the three main subsystems. (C) is a version-labelling block. It makes a firmware version number available to be read from software while the firmware is running. (I) is some logic that generates a signal indicating whether the FPGA initiated reading out data from the whole array (see section 3.6). (H) is logic to select whether the data source to the firmware is from the ADC data input from the main design or one of several test signals. The test signals include a constant signal, a 10-bit counter, and a long-running counter. The selection is controlled by a software register. It is important to note that additional test signals are available from the main firmware, and so for the cosmic ray system to receive data from the ADCs, the ADC data source must be selected here and also in the main firmware settings. Within the cosmic ray design, choosing the constant test signal replaces the timeseries corresponding to each antenna with a different constant value for each antenna, corresponding to the antenna number. This test signal is useful for tests such as verifying that data from each antenna is received in the Ethernet packets. This constant signal is a 10 bit constant for each antenna, because the ADCs output 10 bit data and the test signals must be the same bit width. The 10-bit counter replaces the timeseries from all antennas with an identical signal that counts from zero to 1023, incrementing by one on each clock cycle and wrapping back to zero after reaching 1023. With the long running counter, each antenna's signal is replaced with 10 bits of a 40-bit counter, such that groups of four antennas contain all the information from the counter. Combinations of tests using the counters verify the data format as well as helping test the trigger and veto subsystem.

The cosmic ray firmware as a subsystem of the whole LWA firmware

The entire cosmic ray firmware is a subsystem within the main LWA firmware design. The main design also contains the filterbank firmware that computes frequency-domain channels for the other astronomy observations (see Figure 3.3). For the OVRO-LWA upgrade, Jack Hickish developed the filterbank firmware that sends data to the correlator or beamformer data recorders, and I developed the cosmic ray firmware. Since the FPGAs can have one configuration at any given time, building a system in which all these observing modes will be able to function commensally requires compiling all this firmware into one design. To manage this unification, the cosmic ray system block exists in a separate simulink file, which can compile and run firmware simulations faster than compiling or simulating the entire design. The cosmic ray subsystem block then is copied into the main Simulink file to compile the whole design, a process which could be streamlined in the future by incorporating the cosmic ray subsystem into a library to import. The cosmic ray block must be compiled with the main OVRO-LWA firmware design in order to access ADC data, synchronized time stamps, and antenna-based cable delays. The separate cosmic-ray-only Simulink file does compile, generating firmware that can be used for tests of certain parts of the system (such as Ethernet and inter-FGPA communication).

A few notes about the documentation in this chapter

In documenting the firmware, "antenna" refers to the digitized voltage timeseries (and timeseries of other quantities derived from it) from a single dipole, and the fact that the 64 antennas per SNAP2 board correspond to 32 dual-polarization pairs does not become relevant until the next chapter. The firmware treats all antenna signals the same except for their individual cable delays and potential role as a veto antenna. All the arithmetic is fixed-point (as opposed to floating point) arithmetic. Unlike in many software programming languages, the firmware-compilation software handles arbitrary bitwidth numbers up to some large maximum limit in Simulink, and I allow arithmetic with full-precision (i.e. the sum of two 8-bit input numbers produces an output 9-bit number) wherever feasible.

In the firmware documentation in this chapter, I discuss all the main features, but

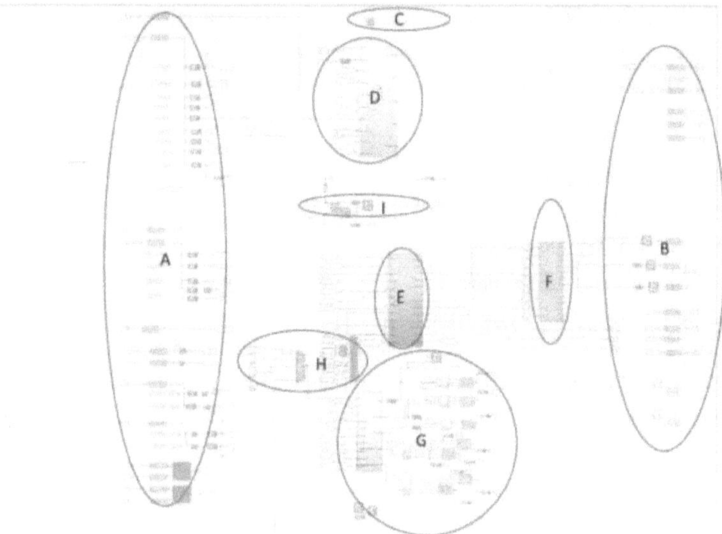

Figure 3.1: Annotated Simulink diagram of the top level of the cosmic ray system. The details of each block are viewable interactively in the Simulink file linked to this book in the Caltech book archive. The circled sections of the design are: (A) Inputs from software registers and general-purpose-input-output pins, (B) outputs to software registers and general-purpose-input-output pins, (C) version label block, (D) state control subsystem, (E) buffer a nd p acketizer s ubsystem, (F) CASPER 40 Gb Ethernet block, (G) trigger and veto subsystem, (H) data source selection, and (I) metadata calculation to indicate whether the FPGA sent a trigger signal.

I don't show figures for the full hierarchy within blocks that consist of many lower level components, and I don't discuss each of the delay blocks needed to keep different signals s ynchronized. In fi gures wi th simplified dia grams, the simplified flow charts use lines to show how information is exchanged between different parts of the design but do not generally have a separate line for each input and output port of each block. To explore the design interactively, the Simulink file will be linked to this book in the Caltech archive and is also available on Github[2].

3.3 Overview of the inputs and outputs to the cosmic ray firmware

There are three sources of input to the cosmic ray system: run-time input from control software (via software registers), signals from other parts of the firmware design (via FPGA interconnect), and signals from other SNAP2 boards (via general purpose input output pins). The types of outputs are software registers, shared BRAM, GPIO pins and 40 Gb Ethernet. The values in the software registers are all readable from software. The registers that act as inputs to the cosmic ray system firmware are writeable from software, and the registers that act as outputs to the cosmic ray system are writeable from firmware. Shared BRAM functions similarly as output software registers, except that Shared BRAM is configured to a chosen size and number of addresses. The software registers and shared BRAM are described in more detail in the section on software control. The general purpose input output pins are used for signaling directly between SNAP2 boards (see section 3.4). The 40 Gb Ethernet output transmits snapshots of timeseries data off the SNAP2 boards. The inputs from the main firmware design (see Figure 3.3) are as follows:

1. **sync** is a Boolean integer that forms a synchronization pulse that allows blocks which need a synchronized time reference to start a counter at the time that sync=1.

2. **din** is an unsigned 640 bit integer that busses together all 64 10-bit data signals.

3. **tt** is a 64-bit unsigned integer that equals the Unix time, in number of clock cycles, at which the SNAP2 last synchronized its timekeeping blocks. During the synchronization procedure the SNAP2 board receives the tt value which it should set its timekeeping block to output on the next sync pulse.

3.4 State Control

The cosmic ray firmware operates in two states: reading out buffered data or recording new data. In this section I present the logic components that manage whether the board is reading out data, which I refer to as "readout state" or recording new data, which I refer to as the "listening" state. Three types of condition can cause a SNAP2 board to transition from listening state to readout state. (1) The logic searching the data for candidate events can detect an event that meets its trigger criteria (see section 3.5). (2) The SNAP2 can receive a signal from its software interface indicating that it should read out the buffer. This feature is useful for

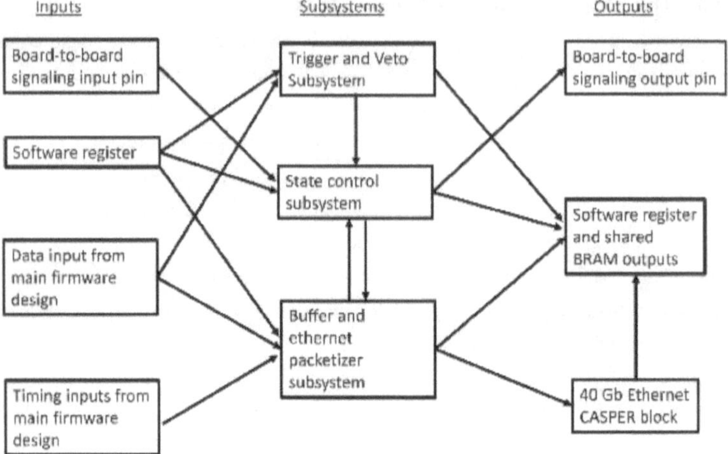

Figure 3.2: Flowchart summarizing the types of inputs and outputs to the three subsystems of the cosmic ray portion of the OVRO-LWA firmware.

testing purposes and also for sampling the background noise conditions by allowing the buffer to be read out at times without any events. (3) The SNAP2 board can receive a trigger signal from another SNAP2 board. This feature makes it possible to save data from every antenna, including from SNAP2 boards whose antennas' signals did not meet the trigger criteria, in order to search for sub-threshold pulses at the edges of a cosmic ray footprint. The firmware state changes from "readout" to "listening" when readout completes or when the board receives a forced reset signal from a software register.

The complete list of inputs to the state control logic is:

1. **trigger** is a Boolean signal indicating that system should switch from listening to readout, if coincidence trigger is enabled. It is driven by the coincidence trigger logic.

2. **readout_done** is a Boolean signal that equals one when readout completes, indicating that system should switch from readout to listening state. It is driven by the buffer and packetizer logic.

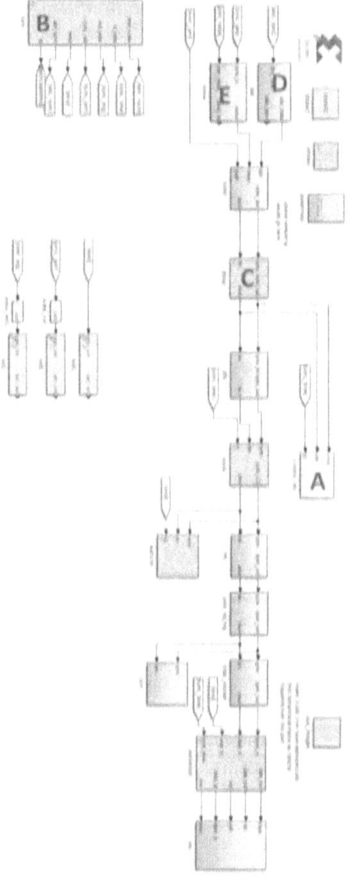

Figure 3.3: Top level Simulink diagram of the OVRO-LWA firmware, showing the cosmic ray subsystem (A) within the main firmware design. Blocks relevant to the cosmic ray detector are labelled with blue letters. The synchronization block (B) produces a synchronization signal and a timestamp (Unix time in number of clock cycles) which are input to the cosmic ray system as the "sync" and "tt" inputs, for the purpose of timestamping the data. The timestamp is input directly to "tt", and the synchronization signal passes through the upstream blocks first. The delay block (C) applies delays to the data to correct for each signal path's different cable and electronic delays. The ADC block (D) handles the interface between the FPGA and the ADC, and outputs the raw timeseries of ADC samples. The noise block (E) offers the option of replacing the ADC data with simulated signals for testing purposes. After passing through the data source selection and the delay block, the data signal splits and one copy goes to the "din" input of the cosmic ray system while the other copy proceeds to the filterbank logic.

3. **rst_to_listen** is a Boolean signal that, on changing from zero to one, indicates that the system should change from readout state back to listening state. It is controlled by a software register. This feature was only useful during testing the Ethernet readout.

4. **en_coinc_trig** is a Boolean signal indicating whether trigger signals from the coincidence trigger should be accepted (value of 1) or ignored (value of zero). This signal is important so that triggers don't flood the system while setting up to run the trigger, and also as a precaution in case some boards do not have enough working veto antennas.

5. **wait_after_read_done** is an Unsigned 12 bit integer indicating a number of clock cycles to wait after readout completes before a trigger from the coincidence trigger logic can trigger another readout. When one board triggers on a candidate event, all the boards read out their buffers. It takes a finite amount of time for the trigger signal to travel to all of the boards, and so they never start and finish readout at exactly the same time. During the wait time specified by wait_after_readout, the board will only start another readout if it receives a trigger from another board or from software, but its own internal coincidence trigger signal is ignored. The value of wait_after_read_done set from a software register is added to a fixed wait of 4096 clock cycles to allow the buffer to fill with new data. This wait time value is controlled by a software register

6. **software_trigger** is a Boolean signal that, on changing from zero to one, indicates that the board should read out its buffer. It is controlled by a software register.

7. **gpio_in** is a boolean signal that, on changing from low to high, triggers the system to read out the buffer. It is driven by a general purpose input output (GPIO) pin which is linked via a wire to a neighboring board, and is of a type such that the low voltage is 0 V and 3.3 V is high.

8. **trig_debug_reset** is a Boolean signal that, on changing from zero to one, resets a counter that counts the number of triggers received from all sources. It is controlled by a software register.

9. **trig_timer_reset** is a Boolean signal that, on changing from zero to one, resets a counter that counts the number of clock cycles between consecutive triggers. It is controlled by a software register.

10. **snapshot_count_reset** is a Boolean signal that, on changing from zero to one, resets a counter that counts the number of times the board changes from readout state to listening state. It is controlled by a software register.

The complete list of outputs to the state control logic is:

1. **readout_state** is a Boolean indicating whether the board is in the listening state (0) or readout state (1).

2. **out_gpio** is a Boolean signal that changes from zero to one (and holds the one for 1000 clock cycles) to send a trigger signal to a neighboring board. It drives a GPIO pin which is wired to the neighboring board.

3. **trigger_count** is a 32bit integer, corresponding to the number of triggers the board has received from all sources in the time since the counter was last reset by a pulse on the trig_debug_reset port. It is sent to a software register

4. **trig_timer** is a 32b bit integer counting the number of clock cycles between two consecutive triggers, unless reset by a pulse on the trig_timer_reset port. It is sent to a software register

5. **snapshot_counter** is a 32bit integer, counting the number of readouts since the last time the counter was reset by a pulse on the snapshot_count_reset port. It is sent to a software register.

Figure 3.4 shows the main section of the state control logic. The part of the subsystem not shown here monitors outputs and is shown in Figure 3.5. The logic to determine the state consists of logic to process the input trigger and reset conditions, followed by a counter and a relational that checks whether the output of the counter is greater than zero. The Boolean output of this relational is the readout versus listening state. One means readout state and zero means listening state. The counter is reset to zero by either a readout done signal or a software reset. Triggers increment the counter, but the state is only determined by whether the counter value is zero or greater than zero, such that repeat triggers are ignored. The readout state is output to the rest of the system. When the readout state changes from zero to one, a trigger signal is sent to the GPIO pin. Before sending the trigger to the GPIO pin, the pulse signal (value of one for a single clock cycle) is extended in order to prevent ringing at the pin. If the FPGA is already in readout state, any subsequent triggers received are not passed on the the GPIO pin, in order to prevent triggers passing from board to board in an unending loop.

The logic that processes the input trigger conditions checks whether the required wait time has passed after a readout. This wait time consists of the sum of the software-controlled time to wait to account for the delays for different boards to finish readout and 4096 clock cycle wait for the buffers to have filled with new data. A comparator also checks whether the software register bit to disable the board's event search has been set.

The logic that processes the conditions for changing from readout to listening state consists of a simple logical or that combines the reset signal from software with the readout_done signal from the buffer-and-packetizer subsystem.

The remainder of the logic in the state control subsystem consists of counters to record various conditions and make these statistics available to output software registers, shown in Figure 3.5. A counter acts as a timer counting the number of clock cycles that elapse between two consecutive triggers and then making the count available to a software register. After two triggers, it does not continue timing triggers until it has been reset. The register is useful for timing the number of clock cycles that it takes for a trigger signal to pass to all the other boards. The other diagnostic information available in this section is a counter that counts the number of times that the board has changed from readout state back to listening state. Unless the forced software reset has been used, this count is the number of times the buffer has been read out. The counter can be reset from software.

3.5 Coincident Event Trigger and Anti-coincidence veto subsystem

The trigger and veto (also referred to as pulse search) subsystem searches the data for events that illuminate a threshold number of antennas from the core array and do not illuminate the distant, veto antennas. This subsystem is the first step of the cosmic ray search and contains all the on-FPGA RFI flagging.

Inputs and Ouputs

The inputs to the trigger and veto subsystem are:

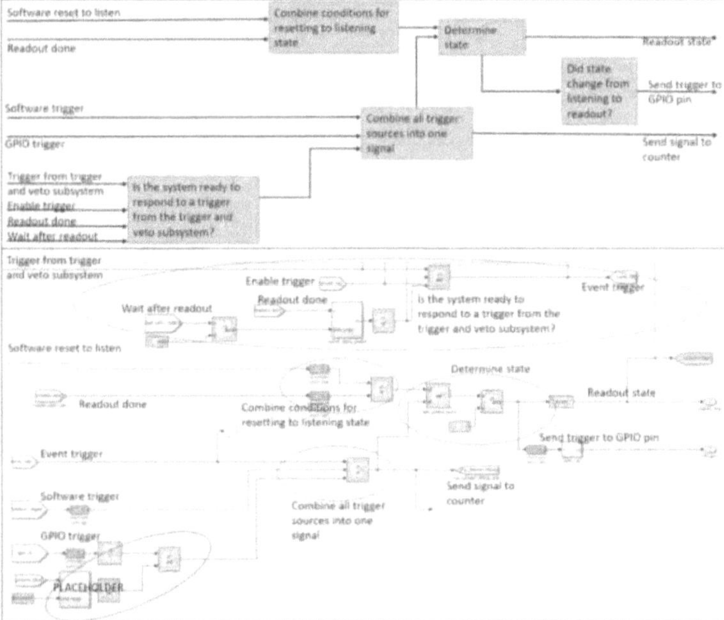

Figure 3.4: Block diagrams of the main part of the state control firmware. Top: Simplified diagram. Bottom: Annotated Simulink diagram. For clarity, the logic near the top of the Simulink diagram that determines whether to respond to a trigger from the trigger and veto subsystem is moved to the lower left in the simplified diagram, to avoid breaking lines representing signals. The logic labelled "Placeholder" does not appear in the simplified diagram because it is redundant and will be removed in a future version of the firmware.

1. **data_in** is a 640-bit bus of input data that concatenates 10 bit unsigned integers for all 64 antennas.

2. **sync_in** is a Boolean place holder for a synchronization signal which has not been needed.

3. **ant_thresh_in** is a signed 32-bit integer (signed because that is the default type from the software register, although in this case the value must be positive) that indicates the power threshold that the power timeseries for core antennas must exceed in order to trigger.

4. **veto_ant_thresh_in** is a signed 32-bit integer (signed because that is the

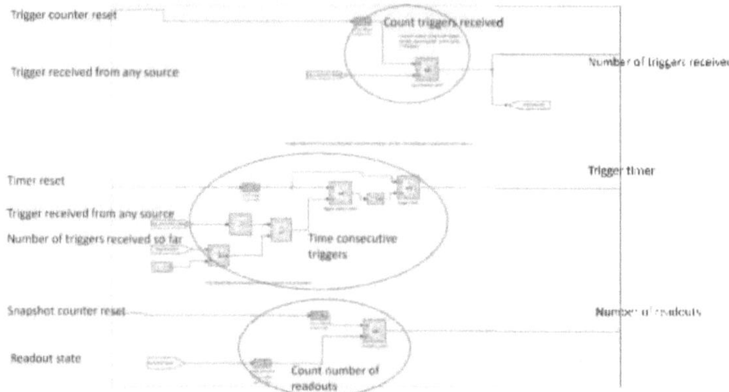

Figure 3.5: Annotated Simulink diagram of the statistics counters in the state control subsystem. Since this section is smaller, only the annotated Simulink diagram is shown.

default type from the software register, although in this case the value must be positive) that indicates the power threshold that the power timeseries for veto antennas must exceed in order to trigger a veto signal.

5. **coinc_window_in** is an unsigned 16-bit integer that represents the window of time (in number of clock cycles) within which to search for multiple core antennas exceeding their power threshold. The window cannot be less than 5 clock cycles for the trigger firmware to function as designed (and realistic windows are much larger than this limit).

6. **veto_window_in** is an unsigned 16-bit integer that represents the window of time (in number of clock cycles) within which to search for multiple veto antennas exceeding their power threshold. The window cannot be less than 5 clock cycles for the trigger firmware to function as designed (and realistic windows are much larger than this limit).

7. **coinc_thresh_in** is an unsigned 7-bit integer. More than this number of core antennas must exceed the power threshold to meet the trigger condition.

8. **veto_thresh_in** is an unsigned 7-bit integer. More than this number of veto antennas must exceed the power threshold to meet the trigger condition.

9. **trigger_antennas_in** is an unsigned 64-bit integer. Each bit represents one antenna and equals one if that antenna participates in the core array.

10. **veto_antennas_in** is an unsigned 64-bit integer. Each bit represents one antenna and equals one if that antenna participates as a veto antenna.

The outputs from the trigger and veto subsystem are:

1. **trigger** is a Boolean which is one when the trigger condition has been met and not vetoed. This triggers the buffer to be read out if the system is in a state that can accept this signal.

2. **core_coincidence** is a Boolean which is one when the trigger condition has been met among the core antennas, regardless of whether there has been a veto. This signal is for event rate logging purposes.

3. **veto_detected** is a Boolean which is one when the trigger condition has been met among the veto antennas, regardless of whether there has been a core trigger. This signal is for event rate logging purposes.

4. **threshold_exceeded_per_antenna** is a 64 bit integer. Each bit corresponds to one antenna and is one if that antenna exceeded the power threshold for the core antennas. This output is for rate logging purposes and is recorded for all antennas regardless of whether they are core antennas.

5. **veto_threshold_exceeded_per_antenna** is a 64 bit integer. Each bit corresponds to one antenna and is one if that antenna exceeded the power threshold for the veto antennas. This output is for rate logging purposes and is recorded for all antennas regardless of whether they are veto antennas.

Trigger and Veto Details

Figure 3.6 diagrams the main parts of the trigger and veto firmware. First, a 24th order finite impulse response bandpass filter with a passband 35-80 MHz filters the timeseries to mitigate RFI at low frequencies and in the FM band. Then, each voltage timeseries is squared to obtain a power timeseries. The resulting power timeseries is smoothed with a four sample moving window average. The smoothing matters because the power in impulsive emission is spread over several samples by the analog and digital bandpass filters.

The smoothed power timeseries is copied and sent in parallel to logic that determines if the core antennas antennas trigger and to logic that determines if the veto antennas trigger. In each case (core and veto), the antenna threshold block outputs a 64-bit number which consists of sixty four Boolean signals bussed together, each indicating whether or not the corresponding antenna exceeded the threshold. The core and veto threshold blocks are identical, but with different threshold inputs in order to allow the possibility of choosing the core and veto thresholds separately. In Figure 3.6, the statistics output from these blocks (Blocks D) is omitted from the simplified diagram in order to capture the main ideas without crossing signal lines. The select_trigger_antennas block sets to zero all streams which do not correspond to antennas participating in the core trigger, and the select_vetoes_block sets to zero all streams which do not correspond to veto antennas.

Next, the outputs of the antenna-selection blocks enter "coincidencer" blocks to determine if above-threshold events coincided among enough antennas to cause a trigger (or veto among the veto antennas). For the core antennas, whenever each antenna's Boolean signal indicating whether it was above threshold rises from zero to one, that Boolean one is extended for the number of clock cycles corresponding to the light travel time across the core array. Then, the extended signals are summed across all antennas, to count the number of antennas that had an above-threshold signal within the light travel time among them. The same procedure occurs in parallel for the veto antennas, but using the light travel time across the entire array. The block labeled trigger_coincidencer and the block labeled veto_coincidencer are the same logic with different window lengths.

The outputs of these coincidencer blocks are compared to the set threshold number of antennas that must detect the event. The comparator is a greater than, not a greater or equal, not a greater or equal, and so the threshold set from software must be exceeded not just met. The minimum number of veto antenna detections to cause a veto is controlled independently from the (in practice larger) minimum number of core antenna detections required to trigger. After these comparator blocks, there is a Boolean signal which is one when the core antennas have detected an event and another Boolean signal which is one when the veto antennas have detected an event.

To account for the time offset between the event arrival time at the core antennas and at the veto antennas, the veto signal is extended for the number of clock cycles equal to twice the veto window, and the core is delayed by number of clock cycles, using a variable delay block. In Figure 3.6 the window input to this block (Block G) is omitted from the simplified diagram in order to capture the main ideas without crossing signal lines. Finally, an AND and a NOT determine whether there has been a trigger that has not been vetoed, and output this trigger signal.

Logic for logging event rates

For all of the outputs from the trigger and veto subsystem, it is useful to measure how often that output is one, i.e. the rate that triggers occur, the rate that core antennas meet the trigger condition, the rate that veto antennas meet the trigger condition, and the rate that each antenna's smoothed power timeseries exceeds the core and veto thresholds. One way to measure these rates would be to have counters that are enabled when the condition occurs and whose output is sent to a software register. Reading the software register twice and dividing by the interval between the reads would give the rate. This is the approach taken for various statistics monitoring in the state control subsystem and the packetizer subsystem. The disadvantage to this approach for the context of rate counting is that reading software registers is an asynchronous process—it is not possible to control exactly how many clock cycles elapse between reads. Instead, for monitoring event rates and threshold excess rates, I use counters that count for a fixed number of clock cycles and then reset. The output of these counters are written to Shared Block RAMS (which are readable from software) on the clock cycle before the reset. Thus, the value read from the Shared Block RAMs is always the count over a known amount of time.

All of the rates I measure are measured by counting ocurrances for 2^{28} clock cycles. I chose this value because using a power of two allows convenient implementation with a Xilinx free-running counter block, and 2^{28} clock cycles is the power of two that gives a count period closest to one second: 1.37 seconds. Figure 3.7 shows the rate counting blocks and shared BRAMs at the outputs of the trigger and veto block. Occurance rates for the following conditions (which I later refer to as events and event rate for brevity):

1. Triggers: The trigger signal sent to the state control subsystem changes from 0 to 1.

2. Buffer readouts: The Readout done signal from the Buffer and packetizer block equals 1.

3. Core duty cycle: The core trigger signal is one

4. Veto duty cycle: The veto signal is one.

5. Core trigger events: The core trigger signal changes from zero to one.

6. Veto events: The veto signal changes from zero to one.

7. Core threshold excess: The filtered, smoothed power timeseries exceeds the threshold set for a core trigger, measured for each antenna (regardless of core or veto status).

8. Veto threshold excess: The filtered, smoothed power timeseries exceeds the threshold set for a veto trigger, measured for each antenna (regardless of core or veto status).

During intense episodes of RFI, an antenna's power time series could in principle exceed the threshold multiple times within the light-travel-time window. This scenario is most significant for the veto, since the light travel time between the veto antennas is multiple microseconds. If a new RFI event arrives before the veto window has ended, the veto window will extend. The veto rate counter that simply counts how often the signal changes from zero to one would not distinguish multiple events. Hence, it is useful to additionally count the total number of clock cycles that the veto blocks events.

The logic for this event-rate-logging is outside the trigger and veto subsystem block because of the CASPER convention of keeping blocks that interface between the FPGA and other hardware (software registers, Ethernet blocks, Shared BRAM, etc.) at the top level of the Simulink diagram (or in this, case, the top level of the cosmic ray system). However, I discuss it in this section because of the functional connection to the trigger and veto subsystem.

Each Boolean data input (1 means there is an event, 0 means no event) passes to the enable port of a counter, such that the output of the counter gives the number of clock cycles that the data input had a value of 1. The counter resets to zero when the period pulse is received. Multiple data streams (in this case, two data streams) are interleaved using the delay and multiplexer blocks. The address output is controlled by a counter that cycles over one address per data input. The address and write enable are timed to cause the counter results to be saved on the last clock cycle before the reset.

Note some of the block names are similar (e.g. count_core_and_veto and count_core_and_veto1) because they are copies of the same block and the names were auto-assigned, even though they are used for different s tatistics. The software for reading these BRAMs will be documented in a software manual accompanying this book in the Caltech library online book archive.

Within the firmware diagram, the periodic pulse generator is labelled "placeholder pps" and the signal is labelled "fake pps", because originally I considered using a pulse-per-second (PPS) SNAP2 signal input to make the counting interval exactly one second, but the synchronization plan for the SNAP2 boards results in only one board having the PPS input connected. Hence, I use a 28-bit free-running counter (see the lower left of Figure 3.7) that wraps back to zero once every 2^{28} clock cycles. A comparator checks whether the counter is zero, and outputs a one on that clock cycle, which is used to reset the rate counters. A future version could replace this with a counter that resets every 196 million clock cycles, for a one-second counting period.

Concluding notes about the trigger and veto firmware

This subsystem treats all 64 input signals identically except for their assigned roles as trigger and veto. Thus, setting the threshold number of veto antennas to be two (i.e. three antennas are the minimum number to detect a veto) means that a veto would occur if three dipoles of the same polarization from three different antenna stands detect the event, or if both dipoles from one antenna stand and at least one from another detect the event. To route the signals through many operations that work on all signals in parallel, sets of 64 signals corresponding to the 64 antennas within the board are concatenated ("bussed") together between operations. Most of the processing operations are implemented with a block that takes an input timeseries of all the antennas bussed together, then splits ("debusses") the input into 64 separate signals within the block, performs its operation on each signal in parallel, and then busses together the output. The diagrams in Figure 3.6 do not show the logic within blocks such as the "coincidencer" block because these consist of a few elements each which I hope can be understood when viewing the Simulink diagram (linked to this book in the online Caltech book Library) alongside this documentation.

Note that many of the logic blocks in the trigger and veto subsystem have a "sync" input and output. These inputs and outputs are connected as a placeholder in case the "sync" signal could be desired for some future feature, but in this design the "sync" signal is not used. In the portion of the subsystem where trigger and veto antennas are processed in parallel, relative delays from the latency of this processing

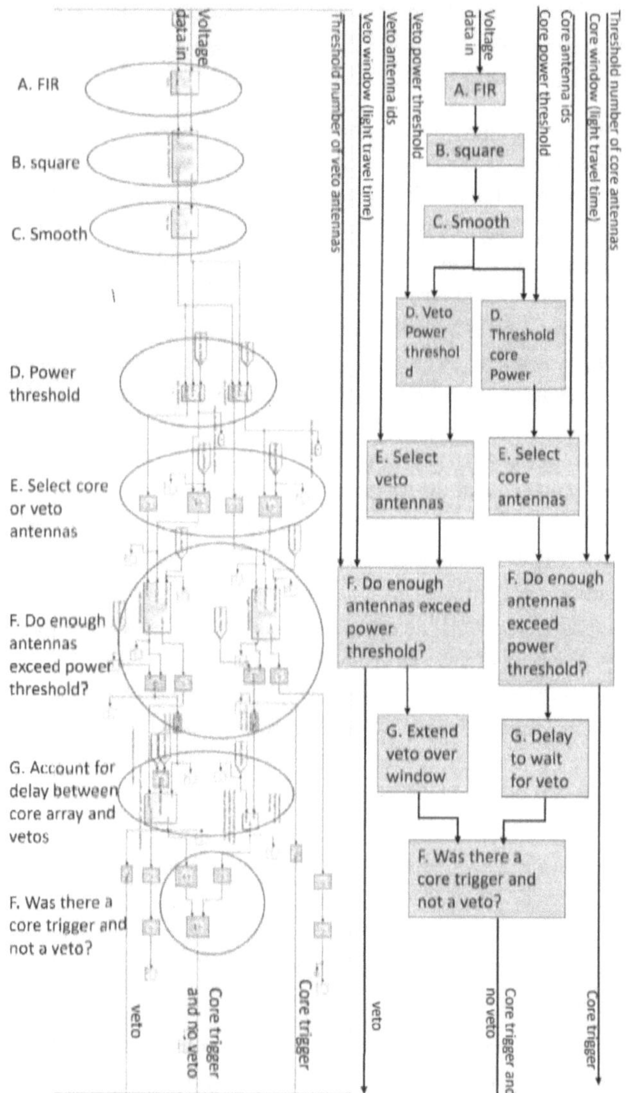

Figure 3.6: Trigger and veto subsystem. Left: Simulink diagram. Right: Simplified. (A) 35-80 MHz FIR. (B) Square voltages. (C) Smooth power timeseries with a four sample average. (D) Threshold on power, output Boolean for each antenna. (E) Select core antennas and veto antennas separately. (F) Count antennas exceeding threshold within window. (G) Extend the veto signal and delay the core to account for all possible relative arrival times between them. (H) Determine whether the core trigger signal is vetoed.

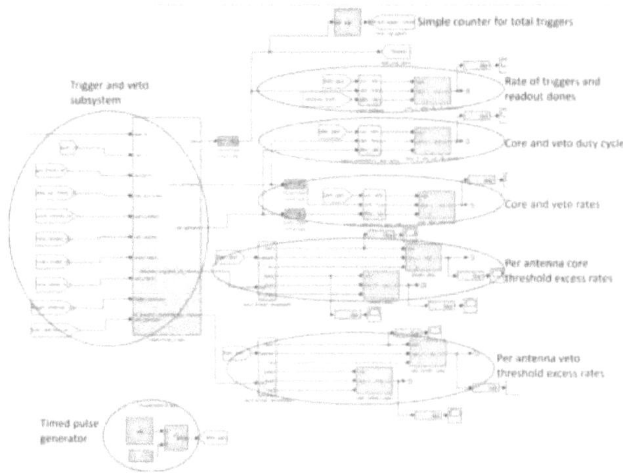

Figure 3.7: Annotated Simulink diagram of event rate logging logic. A simplified diagram is omitted because of the relatively few data processing operations shown. The Shared BRAMs (yellow blocks) are updated once every 2^{28} clock cycles ($\sim$1.37 seconds) with an event count. The top three Share BRAMs have two addresses each and thus each store rates for two different quantities, as indicated in the labels. The lower four BRAMs have 32 addresses each in order to measure the core and veto threshold excess rates for all 64 antennas. The logic labelled timed pulse generator outputs a Boolean 1 once every 2^{28} clock cycles.

matter but are accounted for explicitly by adding delay blocks where needed.

The most complicated part of designing this subsystem was aligning the trigger with the veto. The clock cycles that the veto signal equals True must block *all* of the clock cycles that the trigger is one, not just block any clock cycles. The length of time that the threshold block outputs a one (indicating that more than the minimum number of core or veto antennas detected an event) varies depending on the arrival time delays between the antennas that detect the event, which in turns depend on the arrival direction and location of the antennas. For example, if an event illuminates two veto antennas and arrives from a direction perpendicular to the baseline between them, the coincidencer block will count both of them for the entire length of time that their above-threshold-pulse-signal was extended, whereas if the event arrives parallel to their baseline (maximum delay in arrival times), the coincidencer block may only detect the coincidence for a few clock cycles (one clock

71

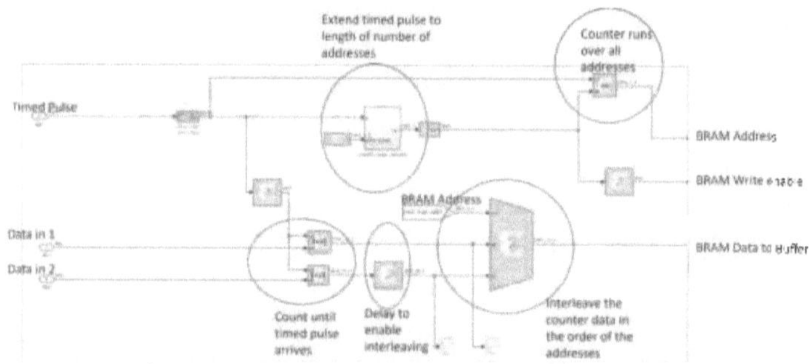

Figure 3.8: Annotated Simulink diagram of an example rate counting block. A simplified diagram is omitted because of the relatively few data processing operations shown. Each Boolean data input (1 means there is an event, 0 means no event) passes to the enable port of a counter, such that the output of the counter gives the number of clock cycles that the data input had a value of 1. The counter resets to zero when the period pulse is received. Multiple data streams (in this case, two data streams) are interleaved using the delay and multiplexer blocks. The address output is controlled by a counter that cycles over one address per data input. The address and write enable are timed to cause the counter results to be saved on the last clock cycle before the reset.

cycle if the coincidence window exactly corresponds to the travel time between the two antennas). Since the length of the Boolean one from the threshold block after the coincidencer block is variable, an edge-detect block is used to turn this signal into a single one for each new time that the number of antennas detecting the event goes from below-threshold to above threshold. Then, that one-clock-cycle value of 1 is the signal that is extended (for veto signals) or delayed (for core triggers). The delay puts the core trigger centered in the window of time that the veto signal (if present) blocks. The total window of time blocked by the veto signal is twice the veto window, which is somewhat longer than strictly necessary. It could be as short as the maximum travel time between a core antenna and a veto antenna, whereas the veto window is set for the maximum travel time between veto antennas on opposite sides of the array. Shortening the time blocked would require an additional software register input. Since the total dead time from the veto is low (see Chapter 4), it hasn't been important to add the extra software control. Furthermore, Monroe et al. [65] found that blocking repetitive events by imposing a dead time after a trigger was a useful RFI mitigation strategy, and in my firmware the veto window can also serve that function.

3.6 Buffer and Data Readout

Introduction to the Buffer and Ethernet Subsystem

The purpose of the buffer and Ethernet subsystem is to continuously buffer 4096 raw ADC samples (~20 microseconds) for all 64 antennas until a trigger condition occurs, and then on receipt of a trigger to transmit that buffered data in UDP Ethernet packets from a dedicated 40-Gb Ethernet port. The interface between the FPGA and the Ethernet hardware is handled by a CASPER 40Gb Ethernet block which is documented in the CASPER repository. The maximum packet size is 256 256-bit words, and thus reading out one 20-microsecond snapshot requires 64 8192-byte packets per SNAP2 board. Conveniently, this packet size allows for one antenna per packet.

Reading out the data takes substantially longer than filling the buffer, for two reasons. The ADCs attached to one SNAP2 board generate 640 bits of data per clock cycle, but the Ethernet block can take in 256 bits of data per clock cycle. Furthermore, data is generated (and enters the buffers) with one sample per antenna per clock cycle, but packets are transmitted one antenna at a time, and the strategy for transposing the data increases the readout time. In an earlier version of the firmware, I achieved a faster readout time by including data from multiple antennas in each packet, but the total readout dead-time is constrained by the rate that the destination computer can receive the packets, and so the multi-antenna packets offered no advantage (and additional parsing software complexity) compared to single-antenna packets with a slower readout. In this section, I document the buffer and packetizer firmware as it is currently implemented. Section 3.9 includes a discussion of design considerations that led to a major change compared to an earlier version of the packetizer.

The majority of the buffer and Ethernet subsystem consists of a subsystem block (named buffer_and_packetizer in the firmware) whose function is to buffer the data and, during readout, to output data in a format suitable for input to the CASPER Ethernet block, along with control signals for that block. The inputs to the buffer and packetizer subsystem block are as follows:

1. **pkt_wait** is an unsigned 32 bit integer that sets the number of clock cycles to wait between sending packets, in addition to the time it takes to form the packet. This wait time is controlled by a software register and used to tune the data rate to avoid overwhelming the network and the destination computer. This value cannot be zero. Attempting to write a zero will not update the

value. For practical purposes, the minimum useful value is larger (roughly 50 clock cycles).

2. **tt** is an unsigned 64 bit integer that, together with sync_in, sets the timestamp in the data packets.

3. **sync_in** is a Boolean integer that, together with tt sets the timestamp in the data packets.

4. **eth40g_rst** is a Boolean integer that resets the Casper Ethernet block if needed for system troubleshooting. It is controlled by a software register and passes directly to the eth40g_rst_out output.

5. **dest_ip** is an unsigned 32 bit integer that contains the IP address of the destination computer. It is set by a software register. The value of this input passes unchanged to the dest_ip_40g output of the buffer and packetizer block in order to be input to the CASPER Ethernet block.

6. **dest_port** is an unsigned 16 bit integer that contains the network port of the destination. It is set by a software register. The value of this input passes unchanged to the dest_port_40g output of the buffer and packetizer block in order to be input to the CASPER Ethernet block.

7. **readout_timer_rst** is a Boolean integer that resets the timer that counts the number of clock cycles spent in readout state. It is controlled by a software register.

8. **data_in** is a 640 bit unsigned integer consisting of the timeseries data for all 64 signals (10 bits each), bussed together as one 640 bit integer. A multiplexer (discussed later) allows a choice between sending raw ADC data as the input (which is the case for all observing) or several different known test signals.

9. **readout_state** is a Boolean integer that indicates whether there is a snapshot to read out (1) or whether the system is ready for a new trigger (0). It is output by the state control subsystem.

10. **eth_enable** is a Boolean integer which controls whether or not the Ethernet packetizer will read out data when triggers are received. This control is important for the sequence of events in which the setup process unfolds as well as for certain tests. It is controlled by a software register.

11. **metadata** is a 256 bit unsigned integer that consists of metadata for each event, described elsewhere in this book. A summary of system settings controlled by software registers as well as a indicator from the trigger and veto system are bussed together to form this input.

12. **delay_buffer** is an unsigned 12 bit integer which is input to prototype logic for an option to delay all the buffer for the purpose of centering events in the buffer. Since triggered events currently fall approximately three quarters of the way through the buffer, which satisfies the need to save samples before and after the impulse, this tunability of the pulse position is not required and the included logic exists as a placeholder for potential future development. It is controlled by a software register.

The outputs of the buffer and packetizer subsystem block are as follows:

1. **eth40g_rst_out** is a Boolean integer that resets the CASPER Ethernet block and is taken directly from the eth40g_rst input of the packetizer subsystem.

2. **data_out** is an unsigned 256-bit integer of data in the desired format for forming packet payloads and is input into the CASPER Ethernet block.

3. **valid** is a Boolean integer that indicates on which cycles data should be added to the packet payload and is input into the CASPER Ethernet block.

4. **dest_ip_40g** is an unsigned 32-bit integer block and containing the destination IP address, taken directly from the dest_ip input of the subsystem and passed to the CASPER Ethernet block.

5. **dest_port_40g** is an unsigned 16 bit integer containing the destination port. Its value is taken directly from the dest_port input of the subsystem and it passes to the corresponding input of the CASPER Ethernet block.

6. **eof** is Boolean integer indicating the end of a packet's dataframe and triggering the sending of the packet. This signal coincides with the last clock cycle for with valid=1 for each packet.

7. **zero_rx_signal, zero_rx_signal1,zero_rx_signal2** are each Boolean integers that are always zero and turn off the unneeded packet-receiving capability. They are inputs to the CASPER Ethernet block that.

8. **eof_counter** is an unsigned 32-bit integer that counts the number of end-of-frame signals (that is, the number of clock cycles where eof=1). This count goes to a software register and provides a measurement of the number of Ethernet packets the SNAP2 board has sent.

9. **eof_and_valid_ctr** is an unsigned 32-bit integer that counts the number of clock cycles for which the valid and end-of-frame signals are both one. This count goes to a software register. It equals eof_counter but is counted separately because it was useful in the process of developing the firmware by confirming the alignment of eof and valid.

10. **readout_timer** is an unsigned 32-bit integer that counts the number of clock cycles spent in the readout state.

11. **readout_done is** is a Boolean integer that signifies when the data readout has completed. This signal goes to the state control subsystem.

All of the inputs to the CASPER Ethernet block are described in the list of outputs from the buffer-and-packetizer subsystem block. All of the outputs of the CASPER Ethernet block are terminated (the design does not require the receving capability) except for a few diagnostics which go to software registers:

1. **led_up** is a Boolean integer signifying whether the link is up, and goes to a software register, bussed together with led_tx.

2. **led_tx** is a Boolean integer that is one for some duration of time related to data transmission and is bussed together with led_up.

3. **tx_afull** is a Boolean integer indicating that the packet payload buffer within the Ethernet block is nearly full. This signal is input to the enable port of a 32-bit counter, and the output of that counter goes to a software register.

4. **tx_overflow** is a Boolean integer indicating that the packet payload buffer has received more than 8192 bytes of data without transmitting a packet. This signal is input to the enable port of a 32-bit counter, and the output of that counter goes to a software register.

Figure 3.10 shows an overview of the contents of the buffer_and_packetizer block. Logic under the heading "Readout timing and control signals" calculates the signals that determine the correct address for reading and writing the buffer BRAM blocks, loops over the different BRAM blocks that make up the buffer, and calculates signals indicating when to bring data into a packet and when to send the finished packet. Logic under the heading "Input signals" manages the inputs described above, and logic under the heading "Generate output signals ready to input to CASPER Ethernet block" converts the timing and control signals internal to this block into signals with the correct synchronization to send to the CASPER Ethernet block as the outputs described above. Logic under the heading "Diagnostics within Simulink only" contains blocks that are used to create simulated oscilloscopes probing different signals in the design, for testing and debugging purposes. These components are only used in Simulink simulations and don't correspond to logic in the compiled firmware. Logic under the heading "Buffer" includes the eight separate BRAM blocks, as well as logic to write data to those BRAMs and to select which block from which to take the data that is read out. The readout process loops over the eight blocks as described later on. The logic labelled "Rearrange data into packet format" performs the necessary transposes to obtain single-antenna packets, and inserts a header. The logic labelled "Calculate packet header" combines the metadata needed to form the packet header, as described below.

Buffer and packet formatting

The voltage timeseries data enters the buffer-and-packetizer subsystem with signals from all 64 antennas (here I treat each dipole as one antenna) bussed together such that the data input is a 640 bit number that updates each clock cycle with the next sample from every antenna. The buffer system must store the 4096 most recent values. This storage is accomplished with Block RAM (BRAM) buffers, but the arrangement of the buffers is carefully chosen to enable a transformation on the output so that each packet, instead of containing one time sample for many antennas, contains all the time samples for one antenna. The data is transmitted off the FPGAs one antenna at a time. Figure 3.12 diagrams this section of the firmware.

The CASPER Ethernet block (shown in the top-level cosmic ray firmware diagram, Figure 3.1) takes in new data as 256-bit "words". The data input port must be a 256-bit integer, and a new 256-bit word can be added to the forming packet once per clock cycle, but need not be added every consecutive clock cycle. A separate Boolean input signal labelled "valid" indicates on which clock cycles the 256-bit integer at the data port should be added to the packet. A packet can contain up to 256 256-bit words, and an end-of-frame "EOF" Boolean signal indicates on which clock cycle the packet's last valid data has entered the block, triggering transmission of the packet. Thus, the key outputs of the buffer-and-packetizer subsystem (see above for a full documentation of inputs and outputs) are a data output of 256 bits per clock cycle, along with a Valid signal and an EOF signal. The Valid and EOF signals must be timed such that adding data to the packet on every clock cycle for which Valid=1 results in a packet containing the voltage timeseries for one antenna.

The simplest way to accomplish the necessary transformation from the input data (1 sample for each antenna on each clock cycle) to the output data (256 bits of data from a single antenna on each clock cycle) is not possible with the address and port options of the Xilinx BRAM blocks. In the conceptually simplest scenario, the data would be clocked into the buffer with 640-bits per clock cycle, and then during readout the data would be read from the buffer in a completely different order, with 256 bits from a single antenna emerging at the BRAM output port on each clock cycle. I did not find a way for Xilinx BRAM blocks to change the address width depending on whether the block is read- or write-enabled. The Xilinx BRAM blocks do allow up to two addresses to be read and/or written on each clock cycle.

My solution uses eight dual-port Xilinx BRAM blocks. The 640-bit data input is de-bussed into eight 80-bit integers (eight antennas each), which go to the 1st data input port of each BRAM. When the firmware is in the "listening" state, this data input is write-enabled and the address port increments on every clock cycle, looping back to address 0 whenever the last address is filled. During the "listening" state, the second set of ports is never used (not write enabled and data input set to zero). When the firmware is in readout state, write-enable is set to zero and the buffer is read-only. One address port of each BRAM loops over all the even addresses and the other loops over the odd addresses, with each address incrementing by two on each clock cycle that the buffers are read out. The two output ports are bussed together into a 160 bit number. The top 80 bits are one time sample for eight antennas, and the lower 80 bits are the next time sample for those eight antennas.

The 160-bit signals from the eight BRAMs enter an eight-port multiplexer block. Multiplexer blocks have a number of data input ports, one "Select" port, and output the input data from the data port corresponding to the value indicated by the "Select" port. In this case the signal at the "Select" port indicates which block of antennas has the antenna whose data is currently being loaded into a packet, and slowly increments over each block of antennas over the course of the data readout. The 160-bit, eight-antenna, two-time-sample, output of this multiplexer then enters a block labelled

"Transpose", which outputs a 160-bit integer consisting of 16 time-samples for one antenna on each clock cycle. To do this, within the transpose block I first re-order the bits of the 160-bit input by de-bussing it into 16 parallel 10-bit data streams that I then bus together in pairs with each pair consisting of two time samples of the same antenna. These eight pairs enter an eight port CASPER transpose block, whose eight output ports on each clock cycle consist of eight successive time samples for a single antenna. More specifically, the inputs over eight clock cycles can be represented as an eight-by-eight matrix whose rows are different antennas and whose columns are different time samples, and the matrix is clocked into the transpose block one column at a time. The output can be represented as the transpose of that input matrix, whose columns are clocked out of the block one clock cycle at a time. Thus, sixteen time samples (because each of the eight signals is already a pair of adjacent time samples) from one antenna appear on one clock cycle, and eight clock cycles later the next 16 time samples for that same antenna appear. The intervening clock cycles present the other seven antennas in that group of eight. The eight outputs of the transpose block are bussed together into a new 160-bit integer for the convenience of creating a single data output port.

Ultimately, taking the data to be valid once every eight clock cycles will provide the entire time series for one antenna. The timing and control signals section creates a preliminary Boolean "Valid" signal which enters the "sync" input port of the transpose block, and is delayed by a number of clock cycles equal to the latency of the transpose, and then is passed to the "sync" ouput port. To cause successive packets to contain successive antennas, the "select_antenna" block takes in the delayed Valid signal from the sync output port of the transpose and delays it by the number of clock cycles (from zero to seven) indicated by the "ant_in_block" input. The timing and control section of the logic ensures that "ant_in_block" increments after every packet, to select the next antenna. The output of the "select_antenna" becomes the "Valid" signal to input to the CASPER Ethernet block, after a few more stages of delays to maintain its synchronization with the appropriate data.

The data output from the transpose block passes to a block that converts it from a bus of sixteen ten-bit numbers to a bus of sixteen sixteen-bit numbers, using zero padding. Sixteen-bit numbers are far more convenient to transmit to the receiving computer, since they are an integer number of bytes. After converting to sixteen bits, the resulting bus of sixteen sixteen-bit numbers is 256 bits, which is the exact size needed for the data input to the CASPER Ethernet block. The final step before sending the data to the data output port of the buffer_and_packetizer block (from which it enters the CASPER Ethernet block) is a multiplexer which selects whether the output should be voltage data or a packet header. One 256-bit word per packet is a header with metadata described below. Since the packets can only be 256 words long, the multiplexer inserts this header at the expense of replacing one word of sixteen time-samples of voltage data. A further 32 time samples from one end of the time series and sixteen timesamples from the other end must be discarded due to an unresolved formatting problem. The final saved timeseries for each antenna is thus 4032 time samples. A delay (the signal "delay_buffer") applied to the data and timestamp inputs to the entire buffer_and_packetizer system ensures that the waveform from the event that triggered the data to be read out always appears roughly three-quarters of the way through the buffer, and thus truncating the timeseries to 4032 time samples poses no significant disadvantage.

Time Stamping and Header Calculation

Time stamping   The logic at the very bottom of the section labelled "Input Signals" is the logic to maintain the telescope time counter, which is used later in time-stamping the packets. This logic follows the approach implemented by Jack Hickish in the filterbank part of the LWA firmware. The 64-bit counter tt_counter has a load port and a data-in ("din") port. On the clock cycle that the load port receives a one instead of a zero, the counter loads the value at the data-in port and commences counting up from that value, incrementing by one on each subsequent clock cycle. The value at the data-in port is the absolute Unix time, expressed in number of clock cycles since the Unix epoch (UTC 00:00:00 on January 1 1970), at which that SNAP2 board was last synchronized (see Chapter 2 for an explanation of the synchronization). The load port receives a 1 when the synchronization pulse reaches it, after being delayed such that the output of the tt_counter corresponds to the Unix time at which the data sample currently entering the buffer-and-packetizer block was sampled by the ADC. In order to keep the time counter synchronized with the data, the output of the counter is subsequently delayed by an amount that corresponds to the delays the data stream experiences passing through the logic it encounters. For example, the timer output and the data are both delayed by an amount that ensures that events that meet the trigger condition end up approximately three-quarters of the way through the buffer. It is important that the event be in a predictable part of the buffer, far from the edges, in order to be able to estimate noise statistics before the event and to check for repetitive RFI afterwards.

Header   After this delay, the timer output enters the logic that organizes the packet header. Every packet includes a header containing the following metadata: a timestamp, which SNAP2 board the data corresponds to, which antenna the packet corresponds to, whether the data is being read out due to a trigger that originated within the same SNAP2 board, and several parameters describing the current settings of the trigger and veto subsystem: the power threshold for core antennas to cause a trigger, the power threshold for a the veto antennas to veto the trigger, the threshold number of core antennas that must detect the event to cause a trigger, the threshold number of veto antennas that must detect the event for it to be vetoed, the coincidence window for core antennas, the coincidence window for veto antennas, and whether the antenna for that packet was a veto or core trigger antenna. All of the above information except the timestamp and antenna number originates outside the buffer and packetizer block and is packed into a 256-bit header with minimal processing. The antenna roles (which antennas are core antennas and which are vetos) enter the block with two bits per antenna. One of the bits indicates if the antenna participates in the core trigger, the other indicates if the antenna participates as a veto antenna, and both bits are needed because malfunctioning antennas are set to have neither role. The packet header does not have room for these 128 bits, and so these bits are split off from the rest of the header, split into individual single-bit data streams (henceforth referred to as "de-bussing") and then a series of multiplexer blocks selects only the bits corresponding to the veto and trigger status of the specific antenna in the current packet. These bits are then concatenenated ("bussed") into the rest of the header. Some zero-padding is included to make the header occupy the required 256 bits exactly. The signal containing the antenna number comes from the timing and control signals logic within the packetizer.

Calculating the timestamp for the header requires slightly more logic. The telescope time counter output is buffered similarly to the data, with a buffer of the same length that, like the data buffers, is write-enabled whenever the firmware is in the "listening" state. The buffer address input is a simple counter that increments whenever the address is write enabled. Unlike the data buffers, however, which must cycle through all address values during readout, the timestamp should be a fixed time that is the same for every Ethernet packet that belongs to one snapshot from one board. Thus, during readout the address does not increment and remains fixed at the value the address took when the write-enable went from 1 to 0. The relative time that a cosmic ray signal arrives at different antennas matters for identifying and reconstructing air showers, and should be known to the clock cycle, but the absolute arrival time is less crucial. The packet timestamp corresponds to the sample time of data in the same buffer position for every packet (and thus every antenna), but this system tolerates an unspecified fixed offset between the timestamp and the start of the buffer. The packet timestamp concatenated with the rest of the packet header forms a 256-bit header word. The header output only changes between packets, and so the output is delayed by 128 clock cycles to more than compensate for the time that the data takes to pass through the logic that formats it into packets. Figure 3.11 diagrams the logic for the header.

One byte in the header indicates whether the trigger signal to read out the buffer originated within this SNAP2 board (as opposed to being received from a neighboring board or from software). The logic to prepare this bit for inclusion in the header is located outside the buffer and packetizer subsystem block, between that subsystem and the state control block, and is shown in detail in Figure 3.9. The signal to indicate that the trigger originated within this SNAP2 is the output of a counter which receives a one at the enable port if a trigger from the trigger and veto subsystem occurs when the board is not already in readout state. This counter is reset to zero when the readout state changes from one to zero. The counter is eight bits mainly because that is a convenient size to add to the header, although a Boolean would suffice for the information of interest. It is technically possible for multiple trigger signals to occur in the time it takes for the readout state to change (although I have not sought to search for this phenomenon), and in that event this byte in the header would report the number of such triggers. It is also technically possible for one event to have packets from multiple boards with a non-zero value for this header byte, since if a large number of antennas are illuminated then multiple boards could independently trigger on the event faster than the trigger signal travels to all the boards. If a board's trigger and veto subsystem triggers after the board is already in readout state (due to a neighboring board having triggered first), then the trigger would not be logged in this byte.

Timing and Control Signals

Here I describe how the various signals are generated that control the timing of activity in the packetizer. The scheme described in section 3.6 for reading out the buffer requires the following signals:

1. **block_step** is a three-bit integer selecting which of the eight groups of eight antennas contains the antenna whose data is currently being read out into

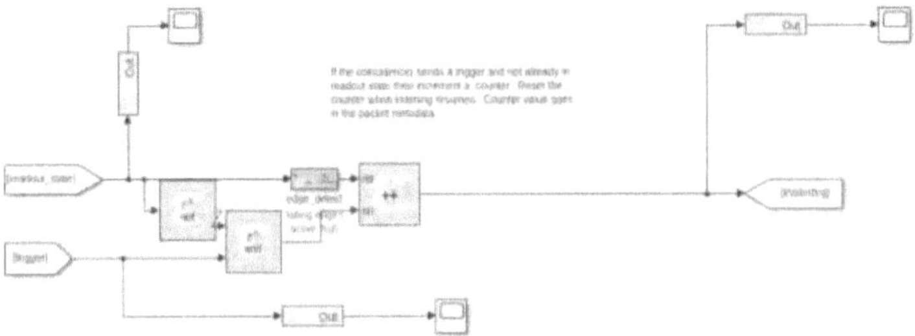

Figure 3.9: Simulink diagram of the logic used to calculate the bit of header metadata that indicates whether the SNAP2 board sending the packet originated the trigger. This logic is located outside the buffer and packetizer subsystem block, between that subsystem and the state control block.

packets.

2. **ant_in_block** is a three-bit integer selecting which of the eight antennas in the group is being read out into packets.

3. **buffer_addr** is the address to be read or written in the BRAMs. The same address is copied to all eight BRAMs.

4. **buffer_we** is a Boolean that equals one if new data will be written to the buffer BRAMs at the address indicated by buffer_addr and zero if data at the current address should be read only.

5. **valid** is a Boolean that equals one once every eight clock cycles during data readout, to indicate that good data should be added to the packet. The Valid output by this section of the logic is not directly input to the Ethernet block, but undergoes some transformations described in 3.6 to produce that input to the Ethernet block.

6. **readout_done** is a signal indicating the end of the readout process, which is sent back to the overall state control logic in order to resume the firmware state in which new data is written to the buffer while the system waits for a new trigger.

The logic to calculate the above signals (described in Figure 3.13) takes two input signals from outside the buffer_and_packetizer subsystem:

1. **readout_state** is a Boolean that is one when the firmware is in the process of reading out the buffer and is zero when the firmware is recording data and waiting for a new trigger.

2. **packet_wait** is the number of clock cycles to wait between sending subsequent packets, in addition to the number of clock cycles required to form the packets. This signal ultimately comes from a software register, allowing the packet rate to be tuned as part of configuring the packetizer. The readout system requires this wait time because the FPGAs are capable of sending packets faster than the network can handle.

Writing to the buffer is enabled whenever not in readout state (see A in Figure 3.13). The counter shown as (B) in Figure 3.13 determines the buffer address that the BRAMs are writing to when the board is not in readout state.

The overall timing of the process of reading out each packet is timed by the counter "count_limited_counter", labelled (C) in Figure 3.13. The overall process is to wait for the software-tunable time between packets and then read out the buffer by cycling through addresses in steps of two, with odd and even being read out on the two different ports. The overall packet cycle counter counts to the sum of 2048 and the chosen packet wait. It is 2048 and not 4096 because the address will increment in steps of two and both ports will be used to read out the buffer twice as fast as it was written, such that cycling through all 4096 buffer addresses once takes only 2048 clock cycles. When the counter reaches the sum of 2048 and the packet wait time, it begins counting from zero again as long as there is a Boolean 1 at the enable port and a Boolean 0 at the reset port. It is enabled whenever the board is in readout state and reset at the end of the readout. The reset is needed because it takes some clock cycles for the readout_done signal to go through the state control logic and result in a change from readout state to listening state, and it is important that the packet cycle counter not continue counting during this time but stay at zero to be ready for the next readout.

The antenna counter (D) in Figure 3.13 keeps track of which of the 64 antennas is currently being read out. It increments whenever the main packet cycle counter resumes incrementing after falling back to zero. The upper three bits of the antenna number determine which group of eight antennas has the antenna being readout currently, and the lower three bits determine which antenna within that group is being read out (see (E) in Figure 3.13). The antenna number, and thus these two indices, update during the pause between packets, making the synchronization of these signals less strict.

After the antenna counter reaches 63, the next time it increments it wraps back to zero, indicating the end of the readout process (see (F) in Figure 3.13). The Boolean readout_done goes back to the state control to trigger changing from the readout to the listening state. This signal also generates a reset signal which resets all the counters involved in the reset process. The only difference between the signal "readout_done" and "readout_done_rst" in the firmware is that the latter (the reset signal) extends the readout done pulse for enough clock cycles to hold the counters at zero until all the signals have propagated through the system. Note that the last packet will be sent some (~ 16) clock cycles after the readout done signal goes to 1, because it takes some time for the last data that is already on the way to make it through the packetizer.

While the overall packet cycle counter is counting, data is only read out after the desired wait time between packets has elapsed, as determined by the comparator at (G) in Figure 3.13.

The logic at (H) generates a Boolean signal that equals 1 once every eight clock cycles when data is being read out (i.e. this counter is not enabled during the wait time between packets and during the listening state), in order to select which clock cycles of the data stream out of the transpose block have data from the same antenna.

The logic at (I) calculates the buffer address during readout with a counter that increments whenever not in the listening state and not during the wait between packets. The counter counts to 2048 and then the output is multiplied by two in order for addresses to increment from zero to 4096 in steps of two. This value is added to the value of the address at which the buffer address stopped writing, in order for events that cause a trigger to always appear in the same part of the buffer. The multiplexer at (J) determines whether to use the address value calculated for the listening state or the readout state, and outputs the address that is sent to the buffer.

Concluding notes on the buffer and packetizer logic

The only buffer and packetizer logic not shown in Figures (3.10-3.13) are some counters at the outputs to the block, which generate useful diagnostic information that can be accessed by software registers. These diagnostics include the number of end-of-frame signals, the number of end-of-frame signals coinciding with valids (this count and the latter are the same, but verifying that they are the same was an important test during debugging), and the number of clock cycles that elapsed during the last readout. I have not detailed the reasons for each of the delay blocks— they are needed in order to keep all the signals synchronized appropriately. Section 3.9 discusses some further notes on design considerations that led to the design presented here.

3.7 Software

This section briefly summarizes the software for operating the cosmic ray system. The cosmic-ray-specific software will be documented more fully in a manual in the supporting materials linked to this book in the Caltech book archive.

Running the cosmic ray detector requires several setup steps that are general setup for the OVRO-LWA followed by several steps that are specific to the cosmic ray system. Configuring the SNAP2 boards with the appropriate firmware, synchronizing the SNAP2 boards clocks, setting the cable delays, and configuring the analog receiver boards with the appropriate filter and attenuator settings use software developed for all observing modes of the OVRO-LWA. After completing these setup tasks, cosmic-ray specific code sets up the cosmic ray subsystem. This includes configuring the Ethernet packetizer with settings such as the destination addresses and the wait time between packets, and configuring the trigger and veto subsystem with the appropriate thresholds and windows. The full set of steps and options will be detailed in the accompanying manual. Other than running the cosmic ray detector, software exists for tests such as periodically reading out an un-triggered snapshot and running threshold scans where event rates are logged while thresholds are gradually raised.

The control software for the cosmic ray firmware relies on the Casperfpga python library, which handles the software interface with CASPER-programmed FPGAs. The most-used functions are the functions for reading and writing software registers and reading Shared BRAM. The main firmware design includes a small Microblaze CPU, i.e. one corner of the FPGA connects up its logic gates to build a CPU from the FPGA fabric. This Microblaze sends information over the 1 Gb Ethernet interface, making software registers and shared BRAM available to the control software running on a computer on that network, via casperfpga communication functions.

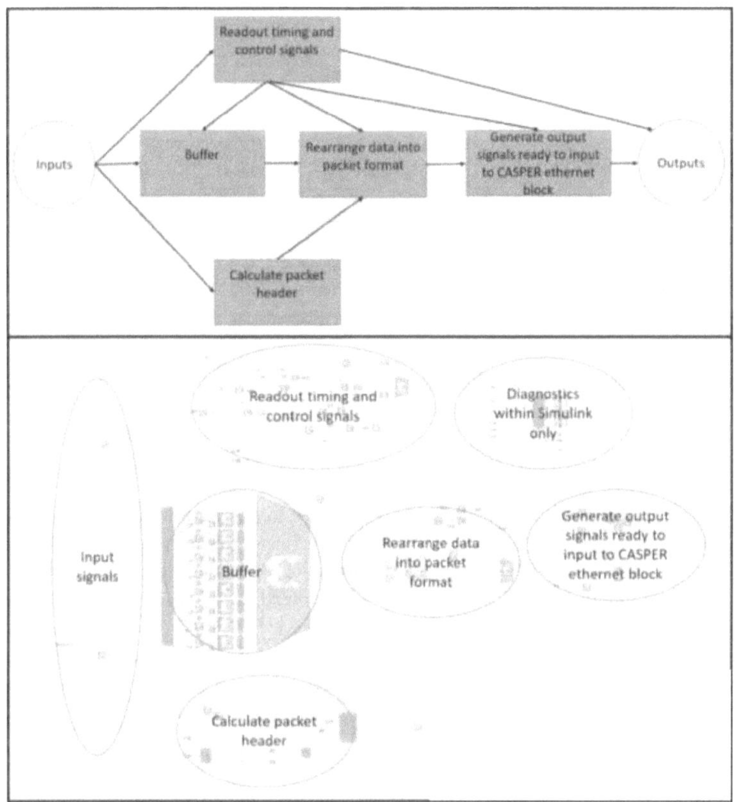

Figure 3.10: Overview diagrams of the buffer and packetizer subsystem. Top: Simplified flowchart. Arrows represent the direction of information flow connecting the main parts of the subsystem. Bottom: Image of the Simulink firmware block diagram with ovals overlaid around the blocks corresponding to the components of the flow chart. Figures 3.11–3.13 show a more detailed view. The group labelled "Diagnostics within Simulink only" contains blocks that are used to create simulated oscilloscopes probing different signals in the design, for testing and debugging purposes. These components are only used in Simulink simulations and don't correspond to logic in the compiled firmware.

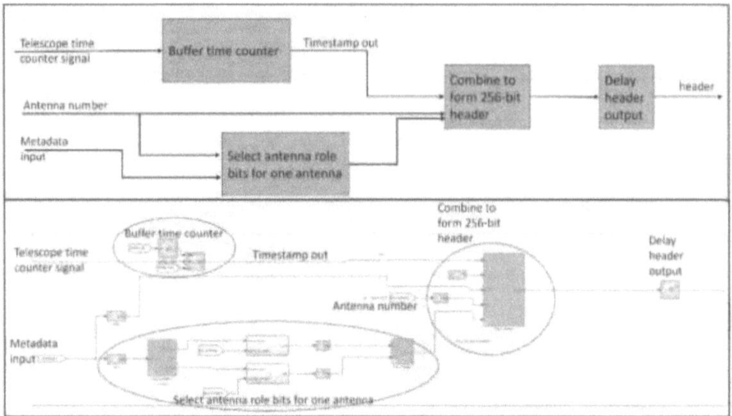

Figure 3.11: Logic for building packet headers. Top: Simplified flowchart. Bottom: Image of simulink block diagram with key components labelled. The timestamp buffer is write-enabled whenever the data buffer is, but instead of reading out the whole timestamp buffer as is done for the data, the timestamp address stays fixed during readout at the place that it stopped writing. Most of the metadata input to the buffer-and-packetizer block passes directly to the final header, but the bits describing antenna roles split off so that only the bits corresponding to the antenna in the current packet go in the header. The antenna id and buffer write-enable signals come from the timing and control logic. The delay at the end is needed to account for the time it takes the data to go through the packetizer logic, so that the header updates between packets.

The data capture code that receives packets from buffer readouts over the 40 Gb network only receives data and does not need to interact with the FPGAs. This code is given options such as the number of packets to put in a file, the total number of packets to receive before stopping, and some destination information, and then simply receives packets and writes them to disk. In the future, a real-time processing pipeline may involve keeping the data in RAM while the first-stages of software processing are performed and then only saving events at a lower rate.

Several network settings and settings within the receiving computer matter for enabling data capture. The 40 Gb network uses static IP addresses rather than DHCP since the firmware needs to be provided with a fixed ARP table when configuring the Ethernet block. On the receiving computer, other than having appropriate addresses and netmask, the network interface (and the 40Gb switch that the data passes through) must be configured to receive a maximum transmission unit (MTU)

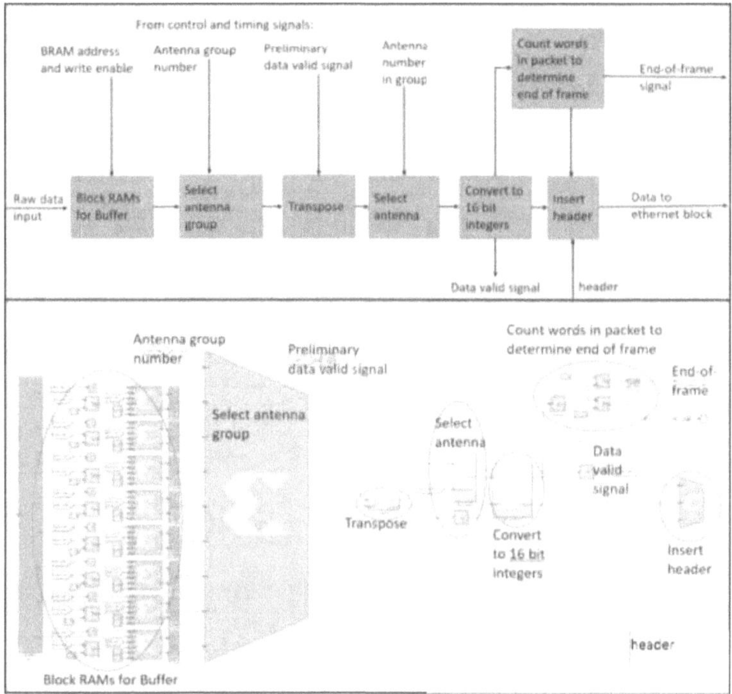

Figure 3.12: Block diagrams of the section of the firmware that buffers the data and formats the buffered data into an order appropriate for single-antenna packets. Top: Simplified diagram. Bottom: Simulink diagram with key parts labelled.

of 9000 Bytes, to be large enough to allow the packets from the SNAP2 boards to be accepted. Additional configuration choices to allow the receiving computer to efficiently capture data from the cosmic ray subsystem at a large data rate include increasing buffer sizes on the receiving computer and seting its CPUs to their high-performance mode. For the Linux Ubuntu 18.04 operating system running on the receiving computer, this configuration can be accomplished with[3]:

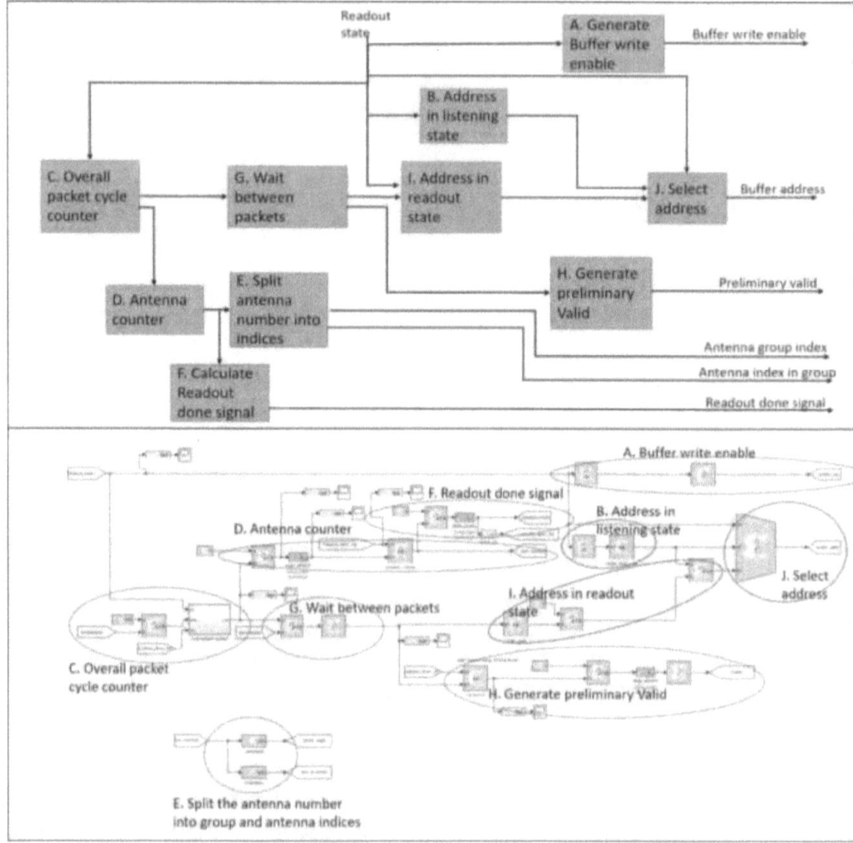

Figure 3.13: Block diagrams of the section of the firmware that controls the timing of the activity of the packetizer system. Top: Simplified diagram. Note that for clarity the antenna counter is portrayed in the lower part of the diagram, avoiding the need for the signal routing tags that the Simulink diagram has. Bottom: Simulink diagram with key parts labeled. See the text for a description of each lettered component.

2. `sysctl net.core.rmem_max=536870912`

3. `sysctl net.core.rmem_default=536870912`

4. `ethtool -G enp216s0 rx 8192`

5. `for i in 'ls /sys/devices/system/cpu/cpu*/cpufreq/scaling_governor';`
   `do echo performance > $i; done`

3.8 Data capture and Network testing

This section summarizes tests to characterize the data rates that the cosmic ray data capture system can handle, with the overall goal of minimizing the detector dead time. The time that it takes for a snapshot of buffered data to be transmitted off the SNAP2 boards in Ethernet packets is time that cannot be used to search new data. The fractional dead-time due to data readout is

$$t_d = r_{\text{trigger}} \times t_{\text{read}}$$

where r_{trigger} is the trigger rate and t_{read} is the time each readout takes. Reducing the readout time reduces the dead time, but making the readout rate too fast risks exceeding the rate at which the data can be received. The tests in this section explore that trade-off, with a goal of under 10% detector dead time at a 50 Hz trigger rate.

There are three distinct rates in question. The **trigger rate** is the rate at which impulsive events occur that trigger a snapshot of buffered data to be read out, for the whole array. The **average data rate** is the product of the trigger rate and the amount of data per snapshot (5.77 megabytes for the entire array). The **burst data rate** is the data rate on the network after one trigger as one snapshot is being read out, and equals the snapshot size divided by the time that the readout takes; the burst data rate $r_{\text{burst}} = 5.77\,\text{MB}/t_{\text{read}}$. The readout time, and thereby the burst data rate, can be adjusted by adjusting the number of clock cycles the SNAP2 board waits between each packet sent.

Figure 2.5 shows a simplified diagram of the network, with SNAP2 boards, a 40 Gb switch, and a receiving computer. Each 40 Gb interface on each device can send or receive data at up to 40 Gb per second. The receiving computer is expected to be able to process less than the maximum rate its network interface card hardware can receive. Based on other systems, the receiving computer is expected to handle 20-25 Gb/s before beginning to drop some data packets. The total data rate arriving at the switch can be much higher than the rate at individual interfaces, because it has many interfaces receiving data from all the SNAP2 boards, but the data rate sent on the single 40 Gb cable to the receiving computer will not exceed 40 Gb.

I tested the burst data rate that I could capture with one 40 Gb port on the receiving computer by sending a trigger signal from software to one SNAP2, which then triggered all the SNAP2s to read out their cosmic ray buffer. I decreased the wait time between packets until the receiving code started saving incomplete snapshots. I tested twelve different packet wait times from 1 to 6000 clock cycles, corresponding to data rates from 68.9 Gb/s to 17.5 Gb/s respectively. Note that the firmware cannot set the packet wait time to zero. The sent data rate can exceed 40 Gb/s because there are 11 SNAP2 boards sending data to the switch and the switch has some buffering capability—the switch will not transmit more than 40 Gb/s to the receiving computer. Before starting the tests I checked that no other processes were using significant computer resources, stopped other code that communicates with the snaps, started the data capture code, and then sent one software trigger to the SNAP2s. Then, I stopped the capture code and recorded the number of packets reported by the capture code, the file size of the output of the code, the network interface statistics from `ifconfig`, and the number of clock cycles that the firmware reports to have spent on reading out the data. I estimated the burst data rate as

$$r_{\text{burst}} = 5.767168 \text{ MB}/t_{\text{read}} = 5.767168 \text{ MB}/(Nr/196\text{MHz})$$

, where Nr is the reported number of clock cycles spent in the readout state. Note that this calculation uses the number of clock cycles that one board spends in readout state. It takes about 167 clock cycles for the trigger to travel around the loop, and so a more accurate estimate of the time over which the data was sent would be Nr+167, but the readout time is always much much longer than 167 clock cycles so that correction is not important. For each packet wait time, I calculated the fractional deadtime that would result for a 50 Hz trigger rate as deadtime=$(50Nr/1.96\text{MHz})/1\text{second}$.

Incomplete snapshots started occurring around 68.2 Gb/s (packet wait times of 20 clock cycles and less), which is almost three times higher than the expected maximum average rate to be received. The burst data rate would exceed the 20-25Gb/s expected limit on the average data rate for readout times faster than 1 ms, corresponding to a packet wait time of 1000 clock cycles. However, these measurements each use an isolated burst of data—one snapshot—and the switch can buffer bursts of data, such that the maximum possible burst data rate can exceed the maximum possible average data rate. The next tests used repeated triggers at a controlled rate.

I repeated the tests decreasing the wait time and measuring the data received using an 11 Hz trigger rate instead of isolated snapshots. 11 Hz is the maximum rate at which I can send software triggers to the SNAP2s due to constraints of the software interface, and faster trigger rates can only be generated with the trigger and veto logic. At the 11 Hz trigger rate, packet loss began to occur (in this case 0.02-0.05% loss) with a packet wait time of 50 clock cycles, corresponding to a data rate during readout of 67.2 Gb/s. The average data rate, however, with this trigger rate is only ~ 0.5 Gb/s (regardless of the wait time between packets).

Testing a faster data rate required internally-generated triggers, using the trigger and veto subsystem with the data source configured to one of the test vector options. I set the data source to the long-running counter which splits a 40 bit counter over the data input for 4 antennas and is duplicated 16 times to provide all 64 data inputs. The penultimate antenna in any set of 4 will have a time series that is equivalent to a 10 bit counter incrementing once step every 1024 clock cycles. I set up the internal trigger on one board and set the second-to-last antenna as a trigger antenna, with no antennas set as vetos. In this version of the firmware the FIR and power smoothing was not applied, so that by setting the core trigger power threshold to 511 and the antenna number threshold to 0 the trigger condition is met once per 2^{20} clock cycles, which gives a trigger rate of 187 Hz. In this way, the appropriate configuration of the test vectors and trigger settings generates a predictable trigger rate that is faster than the rate triggers can be sent from software. To test the data rate that can be received, I ran the data capture code for 10 seconds, measured the number of packets received by the software, and then repeated this test for different wait times. A trigger rate of 187 Hz should give 1321875 packets in 10 seconds of recording, which corresponds to an average data rate of 8.6 Gb/s.

No packets are lost at a 187 MHz trigger rate. Since there is not a suitable test signal compiled in the firmware to test several-hundred-Hertz trigger rates, and since the expected trigger rate is 50 Hz, I did not pursue identifying the exact maximum trigger rate before packet loss would occur. For these last tests, since I had determined that the data rate during readout is a much less strict constraint than the total average data rate, I kept the wait time between packets at 100 Hz which is 2-5 times larger than the value at which packet loss occurs for slow trigger rates, and I now use this packet wait time when operating the cosmic ray system normally.

In the above tests, one SNAP2 board generated a trigger and sent it to the boards. I repeated a test while allowing all SNAP2 boards to trigger simultaneously, which gave a similar result as sending the trigger to a single board did, as expected.

The receiving computer has solid state "Non-volatile memory express" (NVMe) storage drives that can be written much faster than some types of disk. Tests of data capture on a system without NVMes resulted in 4.6%-8.4% packet loss at the 187 Hz trigger rate, even with the longest wait times between packets. With this receiving system, modifying the data capture code to leave the data in RAM resulted in no packet loss, supporting the conclusion that the data loss came from the rate that data could be written to disk. This test shows that the NVMes will be an important part of the system until potentially a real-time software processing pipeline avoids the need to write the snapshot data to disk for every event.

Summary In summary, the wait time between packets, which controls the data rate during read out, is not as important as the average data rate over repeated readouts, as long as it is above ~50 clock cycles. For future tests and observing, I keep the packet wait time at 100 clock cycles. The resulting readout dead time per trigger is 137669 clock cycles, or 0.702 ms. At 50 Hz, this readout dead time corresponds to a fractional dead time of 3.5%, well meeting the goal.

For an isolated trigger, the burst data rate during readout can reach 68.2 Gb/s by relying on the network switch to buffer p ackets. For repeated triggers sent from software at the maximum possible software-trigger rate of 10 Hz, the average data rate is 0.46Gb/s and burst data rates during readout can reach ~67 Gb/s before packets are lost.

The rate that the receiving computer can capture data depends on the type of data storage the computer writes to. Writing to solid state NVMes enabled data capture without packet loss at a trigger rate 187 Hz, (8.6 Gb/s) which was the fastest rate that could be tested with the test signals built into the firmware.

3.9 Discussion

I will conclude this chapter by highlighting a few design considerations, in particular discussing a decision to change the packetizer strategy and arrive at the version presented in this book and currently in use at the OVRO-LWA.

Including tests and diagnostics in the design

Throughout the design and implementation process, I considered where to make information about system conditions available to be read from software registers or

Shared BRAM. Some of these diagnostics have a use in monitoring RFI conditions and optimizing threshold settings (such as the rate counting Shared BRAMs). Others had a specific purpose in testing that the firmware behaves as intended, and do not continue to be used after the version of the firmware passed the tests. For example, the register that measures the number of end-of-frame signals ought to be completely redundant with the register that measures instances of an end-of-frame signal coinciding with Valid=1, since end-of-frame equals 1 on the last clock cycle of the packet for which Valid = 1. However, based on experience with other CASPER Ethernet packetizers, mis-alignments between end-of-frame and valid are a common problem to check for if a firmware version fails to send data, as did occur in one of the early versions of this packetizer firmware.

It was very important in debugging to have access to registers to verify that control signals behaved as expected. Simulating the firmware with Simulink gives access to the value of any signal for every clock cycle for the duration of the simulation, by placing a simulated oscilloscope on that signal. The time it takes to run simulations limits the number of clock cycles for which it is practical to do this type of test. It takes about twenty minutes to simulate one complete snapshot readout, which takes less than a millisecond on the SNAP2s. Thus tests that involve repeated readouts, or running the firmware through a large range conditions must be accomplished with the compiled firmware running on the FPGA. Other than the transmitted Ethernet packets, these tests only have access to software registers and shared BRAMs compiled into the design. There is a limit to how quickly software registers can be read (roughly 1 ms) and so tests running on the FPGA do not have access to what the various diagnostic signals are doing on each clock cycle. For this reason, it is particularly useful to have software registers that take the output of counters that increment when a particular condition occurs.

Making settings adjustable from software and keeping antennas interchangeable

Since compiling and testing the firmware is slow, it is useful to allow important system parameters to be taken from a software register where possible rather than being hard-coded into the firmware. Some parameters, such as the number of antennas and the size of the buffer are a fixed part of the design, but software writes the parameters that set up the trigger and veto system and the Ethernet readout, even for parameters that seldom or never need to change. For example, the time to ignore new triggers after readout finishes is set as the time it takes for the trigger signal to travel to all the SNAP2 boards, since that is the maximum time offset for different boards to start and finish their readout. The length of this delay is determined by the length of the wires, and does not change, but it was useful to design the firmware without needing to know the exact value of that final length, and then set the appropriate value from software rather than hard-coding it into the firmware. Similarly, changing the wait time between packets was only useful in testing the network to determine the appropriate wait time, but leaving this wait time as a value set from software is much simpler than compiling many versions of the firmware with different wait times to test.

All 11 SNAP2 boards run the same firmware and all input data streams from the 64 antennas within each board are handled identically by the firmware. The different antenna roles (trigger, veto, or not participating) are set by multiplying the timeseries with bitmasks set from a software register which is 1 or 0 depending on the antenna's role, but the firmware routes all 64 inputs in exactly the same way. Because of this design, if antennas break or inputs are plugged into different ADC channels, only the bitmask needs to be updated. I considered combining polarization pairs from each antenna stand by summing their power before the thresholding block. I decided to keep the polarizations separate because that greatly simplifies the firmware by keeping all 64 inputs handled identically and because cosmic ray signals are polarized, such that thresholding on the total Stokes I power may not be as sensitive as thresholding on individual polarizations.

One challenge was finding a simple procedure that the firmware on all the SNAP2s perform to make trigger and readout decisions in a way that ensures all SNAP2 boards readout the data when any board's antennas meet the trigger condition. There aren't enough GPIO pins for every board to have a direct connection to every other board, so the boards are wired together in a loop. Each board can receive a trigger signal from one neighbor and send a trigger signal to the other neighbor. It is important that trigger signals don't pass around the loop perpetually, and a simple way to achieve this is that each board passes the trigger on, unless it is already in readout state. Whichever board first sent the trigger does receive a duplicate trigger signal after it makes its way around the whole loop, but the board by then will be in the midst of readout and the duplicate trigger is ignored. The board that first sent the trigger will also finish its readout first, and I needed a way to make sure that it doesn't start another readout before the other boards can respond. But simply adding a minimum time between readouts wouldn't solve the problem because each board would finish its wait time at a different time. The solution is to have a wait time during which the board ignores trigger signals from its own trigger subsystem but would respond to a trigger from a neighboring board.

Ethernet readout performance

The buffer and packetizer subsystem in the current version, which outputs single-antenna packets is a major change from an earlier implementation. The current approach (detailed in section 3.6) transposes the data stream from having one time-sample of many antennas on each clock cycle to have many timesamples of one antenna on each clock cycle. The data is converted from 10-bit to 16-bit numbers before being sent off the FPGA, and 16 timesamples at one end of the buffer are replaced by a header information which includes a timestamp.

The earlier approach theoretically allowed a faster readout time, transmitted all 4096 time samples as 10-bit numbers, and instead of one timestamp sent buffered clock output to time-tag every sample. The advantage of the theoretically faster readout time ended up not useful in practice, because the readout dead time is not limited by how fast data can leave the SNAP2 boards but rather how fast the receiving computer can process the packets. Both packetizer designs require an added wait time between packets. Since there are 11 SNAP2 boards that have hardware to transmit 40Gb/s each, but the data must arrive at one receiving computer, each SNAP2 board should transmit data at most at 1/11th of the rate it could theoretically produce (and in practice the limit is a somewhat lower since the receiving computer must keep up with writing the data).

The first packetizer design sent 64 packets which each contained 256 time samples from 16 antennas, and so each antenna's buffered timeseries was spread across 16 packets. In contrast, the current system truncates the timeseries slightly (it ends up with 4032 usable samples) due to adding the header and a small amount of invalid data at the ends of the packet but doesn't split an antennas data between packets. The earlier version of the packet capture code relied on the order the packets were received to assemble the complete timeseries for each antenna, which were saved as Python Numpy array files. This approach sufficed for testing data capture from a single SNAP2 board but would have greatly complicated the process of organizing data sent from all the boards at once. It could have been possible to assemble the packets using the metadata to identify where in the array the packet belonged, but dropped packets would ruin 16 antennas timeseries at once, and since it had become clear that the receiving computer was the bottleneck, I chose to switch to an approach that was less efficient for the SNAP2 boards but significantly simplified what the packet capture code needed to do.

The receiving code simply writes all the data from each packet to non-volatile storage, and the order in which packets arrive from different antennas doesn't matter because the header identifies the antenna. Typically at any given time a few of the 704 dipoles of the LWA have some problem (e.g. faulty amplifier, loose cable) resulting in a level of data loss which doesn't significantly affect the array's acceptance to cosmic rays. By making one packet per antenna, a missing packet is no more harmful than having an antenna that is down for another reason. Packet loss typically affects fewer than 10% of events (i.e. fewer than 10% of events are missing at least one of the 704 packets, not that 10% of packets are lost). The current version of the event classification software does only operate on complete snapshots of 704 packets, but a planned change is to make the code work with however many packets are received, since a snapshot of 702 packets is still searchable for cosmic rays.

The earlier approach sent the data as 10-bit numbers, which were converted by the packet capture code. The buffer consisted of 4 single-port BRAMs that buffered 16 antennas each, and this 160-bit output required no transpose before being put into packets. The 160-bit data stream was made into the required 256 bits for input to the Ethernet block by adding some header information, including the timestamp on every clock cycle. Sending the entire buffered clock output proved very useful during testing and debugging, but I removed this feature in order to convert the data to 10-bit numbers before putting into packets, which further simplifies what the receiving code needs to do.

With the new packetizer, the minimum possible readout time is 0.67 ms, but to avoid data loss the packet wait time is set to 100 clock cycles, such that the whole readout takes 0.702 ms. The first packetizer had a minimum possible readout time approximately eight times faster but this did not result in an overall lower dead time since the receiving computer, rather than the rate that packets can leave the SNAP2 boards, is the bottleneck.

Conclusion

In summary, the cosmic ray firmware has an event search state and a data readout state. It buffers 20 microseconds of data while searching for events that illuminate core antennas and blocking events that illuminate distant veto antennas. When any boards' antennas meet the trigger conditions all read out their data.

The logic to read out the buffer to Ethernet packets was the most challenging firmware subsystem to develop. In the final version, the readout dead time is ~0.7 ms for each trigger, resulting in a total fractional dead time of 3.5% at a 50 Hz trigger rate. This meets the goal of under 10% fractional dead time by a factor of a few, which gives leeway for the actual trigger rate to sometimes exceed the 50 Hz expectation without excessive detector dead time.

Chapter 4

COMMISSIONING A RADIO-ONLY COSMIC RAY DETECTOR AT THE OVRO-LWA

4.1 Introduction

This chapter discusses the ongoing commissioning of the cosmic ray detection system at the OVRO-LWA. The objectives of the commissioning phase are to determine the configurations of detector settings that best balance the trade-off between cosmic ray sensitivity and RFI-rejection.

This process involves a set of tests and measurements to configure the on-FPGA cosmic ray search (and its associated RFI rejection), as well as the development of RFI rejection software that filters the candidate events sent from the SNAP2 boards.

In addition to triggered snapshots from the cosmic ray firmware, several other data sets inform this process and aid in running the cosmic ray detector. The cosmic ray firmware can read out the buffer in response to a trigger signal from software. I refer to this data as untriggered snapshots, since the snapshot need not contain an above-threshold event. Additionally, the rates at which the chosen thresholds are exceeded can be monitored over the 1 Gb Ethernet interface by reading Shared BRAMs on the FPGAs (see Chapter 3), without necessarily saving the snapshots of data. This capability is useful for experimenting with low thresholds which result in higher trigger rates than the rate at which snapshots can be saved. Finally, some monitoring data made available from outside the cosmic ray system aids as well. Spectra from each antenna, integrated over a longer integration than can be achieved with the cosmic buffer, are saved four times daily for each antenna, using the main filterbank firmware.

The distinct polarization and lateral power distribution function of the radio emission of true cosmic rays will be the definitive features used in identifying cosmic ray air showers among the candidate events saved from the cosmic ray buffer. Thus, the final classification step will be fitting a polarized model lateral distribution function to a map of power received by each antenna. While this fit will be required to identify cosmic rays, most candidate events are RFI, and most RFI can be identified as such with much more simply. Performing LDF fit for every triggered buffer readout would be too computationally intensive to run at a rate that keeps up with

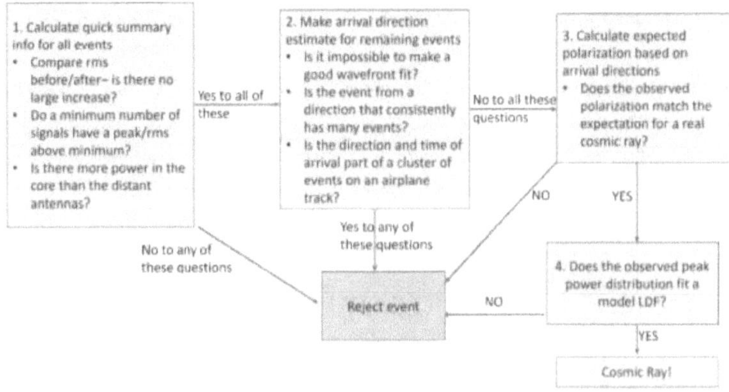

Figure 4.1: Flowchart of potential classification steps to identify cosmic rays among the candidate events.

the influx of new candidates, and so most RFI is rejected on the basis of cuts on statistics that can be computed more quickly for each event. Figure 4.1 shows a flow chart of what a full event classification pipeline might look like.

The first parts of this chapter describe running the cosmic ray detector. Section 4.2 describes the steps for operating the cosmic ray detector (although detailed software documentation will be provided in the supporting materials linked to this book in the online Caltech book archive). Section 4.3 shows the results of reading out the cosmic ray buffer without running the trigger, in order to assess the RFI background. Section 4.4 describes threshold scans and rate logging used to determine suitable configuration settings for running the cosmic ray detector.

As a transition from firmware to software, section 4.5 describes several signal processing decisions, including defining the signal-to-noise ratio used in software and the addition of a finite impulse response bandpass filter to both software and hardware. The remainder of the chapter focuses on developing the fast preliminary cuts, in section 4.6. Section 4.7 concludes with a discussion of next directions for (imminently) future work.

4.2 Setting up the cosmic ray detector for data collection during commissioning

To set up a cosmic ray observation, the first steps are shared by the cosmic ray system and the other astronomy observations. First, the firmware must be loaded. Then the snap2 boards undergo an initialization process including synchronization. Then settings are loaded for the analog receiver boards (filter settings and appropriate attenuation to make the signal power at the digitizers as similar as possible for all antennas). Then the cable delays are set in the firmware, along with some settings that pertain to the filterbank firmware only.

Next, the cosmic ray system can be configured. This involves first configuring the 40 Gb Ethernet ports and testing that the setup succeeded by using a trigger signal from software to send a snapshot from the cosmic ray buffer (with the data capture software running to receive it).

During commissioning, data collection occurs in several-hour observing runs. I perform threshold scans before every data collection run. These scans show for what settings the trigger rate is dominated by thermal noise and for what settings human-made radio frequency interference dominates the rate. These tests are central to the key commissioning objective of balancing cosmic ray sensitivity with RFI-rejection.

After confirming that the event rate appears stable with reasonable threshold settings, I apply the settings and enable the trigger-and-veto system to send triggers to read out the buffer over Ethernet. Separately, the data capture software must be started to begin recording these snapshots.

During commissioning, the software processing does not run in real time, in order to test different event classification cuts by re-processing large saved datasets of candidate events. Thus, instead of running near-continuously as planned for the final system, data collection runs for several hours at night. Night time is chosen for the more stable RFI conditions. Seven terabytes are available on the computer running the data capture software, and thus I run the datacapture with a maximum packet limit set so that it stops recording a little before the drive space is entirely full. After processing the saved data (and typically re-processing to test different cuts), a small amount of useful example events are saved and the rest of the data is deleted to make room for another data collection run.

4.3   Untriggered background snapshots

An early test of the cosmic ray buffer readout used untriggered snapshots read out once per minute for roughly 24 hours. Examining the untriggered background is important to assess whether typical times (not just events that meet the trigger condition) have a background dominated by RFI or thermal noise. These background snapshots show that the OVRO-LWA background is dominated by thermal Galactic synchrotron noise over most of the band, most of the time, but that narrow-band RFI is always present and at times overwhelming (particularly in the daytime). These observations in part motivate the addition of a digital bandpass filter discussed later in this chapter.

Figure 4.2 shows spectra calculated from these time series, for one polarization of one typical antenna. The power spectra were calculated from the square of the fast-fourier-transform of the raw timeseries (with no digital filter applied). The wide darker regions at high and low frequencies show the roll off of the analog filter response. The persistent, narrow bright bands near the bottom of the plot are FM radio.

The time-variable narrow bands across the top of the plot are a variety of low-frequency radio-frequency interference. Unlike the strong FM radio interference, which requires a direct line of sight to the transmitter, low-frequency RFI reflects off the ionosphere, making it observable from sources over the horizon or behind the mountains. During the daytime, the total ionized electron content in the ionosphere increases, which increases the plasma frequency and thus increases the frequency of RFI that can reflect. The ionosphere takes hours after sunset to cool down. Note in Figure 4.2 that during the day time the bands of low-frequency RFI extend to higher frequencies than at night time.

The bright spot near 800 minutes in 4.2 may be the 27 MHz CB radio band [1]. As the solar activity has increased in the year and a half since the data in Figure 4.2 were taken, 27 MHz RFI has increased. The 27 MHz RFI can be strong enough to saturate the ADCs when the standard attenuation settings are used, and has required at times an additional 15 dB of attenuation to prevent the ADCs from saturating. The estimated transmitter power exceeds legal limits, but enforcement is difficult. Due largely to this RFI source, I have opted to initially commission the cosmic ray detector with night-time only data-recording, and later attempt daytime observations.

---

[1] CB radio has been seen in a large amount of OVRO-LWA data since then, although the nature of the source in this particular plot has not been confirmed.

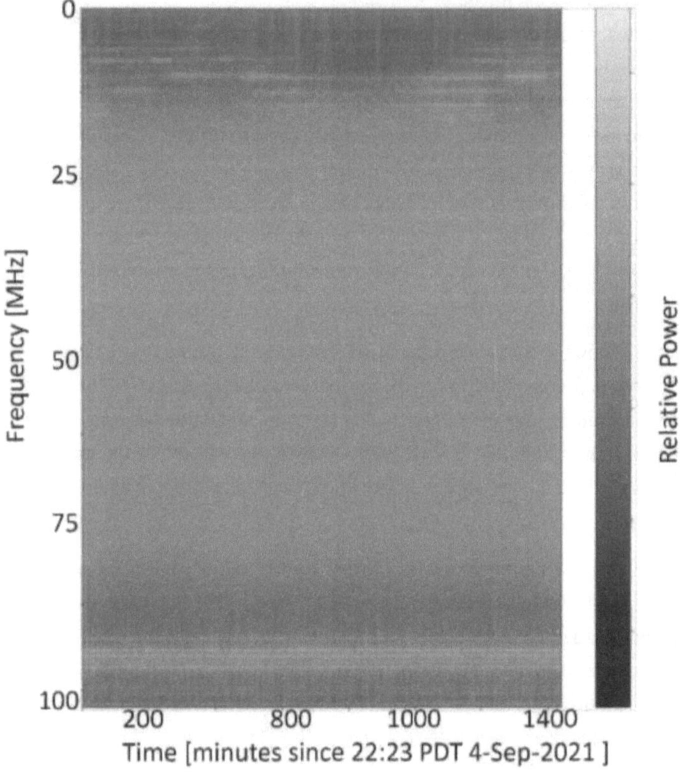

Figure 4.2: Spectra of buffer snapshots read out once per minute over roughly 24 hours. Each column of pixels is a spectrum, with color indicating relative power in arbitrary units. Darker color indicates lower power, and brighter color indicates higher power.

The 27 MHz RFI may be adequately mitigable with the stronger attenuation settings, and the source does disappear for weeks at a time.

4.4 Threshold Scans and Rate Logging

This section describes a set of tests, which provide important information for the commissioning goal of balancing cosmic ray sensitivity with RFI-rejection. I conduct several types of threshold scan. All of these consist of progressively raising or lowering a threshold and counting resulting events. These scans show for what settings the trigger rate is dominated by thermal noise and for what settings human-made radio frequency interference dominates the rate, in order to characterize the RFI background and to help find the optimum settings to run the cosmic ray trigger.

The first is a scan over different voltage thresholds on each antenna within an FPGA. The firmware contains 28-bit counters that increment whenever the signal exceeds the threshold. Counts from all antennas in the FPGA are packed into a shared BRAM where they can be read from software. Every 2^{28} clock cycles, the BRAM is updated and the counter resets (see Chapter 3). Thus, this register provides an estimate of the rate at which the antenna exceeds the threshold, averaged over ~ 1.36 seconds. In the current version of the firmware, thresholding occurs after the raw voltage timeseries is filtered with a finite impulse response filter, squared, and then smoothed with a four-sample moving window average (see section 4.5).

Since two separate thresholds can be set for the trigger and the veto (see Chapter 3), both threshold excess rates are counted and stored in the shared BRAM. For the purposes of a threshold scan, this makes it possible to measure the rates for two thresholds simultaneously. The threshold blocks in the firmware occur before the blocks that apply antenna roles, and so the rates that each antenna's smoothed power timeseries exceeds each threshold is stored in the Shared BRAM. In earlier versions of the firmware, before the FIR was added (see section 4.5), the measured rate was the rate that the square of the raw voltage exceeded the power threshold. This chapter includes results from threshold scans before and after adding the FIR and power smoothing blocks, which will be noted in the relevant figures.

Individual Antenna Threshold Scans

Figure 4.3 shows the results of one per-antenna threshold scan for all the antennas processed by one representative SNAP2 board. Only one SNAP2 board of 64 signals (32-dual polarization antennas) is shown to avoid cluttering the plot. The thresholds on the x axis are in ADC units, i.e. they are a quantity proportional to voltage amplitude and are the square root of the power threshold set in the firmware. The thresholding block in the firmware operates on the smoothed, filtered, power timeseries (see Chapter 3), and the threshold set as input to this block is a power threshold in units of ADC units squared. An amplitude threshold is chosen for plotting because it facilitates comparing the thresholds to the range of the 10-bit ADCs—that is [-511,512]. Note however that the signal processing before the threshold block (apply FIR filter, square, smooth) means that a non-zero excess of a threshold of 511 does not guarantee that the ADCs saturated. Before adding the FIR and power smoothing, it was possible to use the thresholding block to measure the saturation rate exactly.

Most of the antennas in Figure 4.3 show similar behavior. Below an amplitude threshold of roughly 200, the threshold excess rate decreases with increasing threshold very similarly to the expectation for Gaussian-distributed white noise. A threshold scan for white noise would continue to fall off steeply to high thresholds (an early test of the firmware involved running threshold scans after turning off power to the front-end amplifiers to the antennas and decreasing the attenuation so that the input to the ADCs was pure noise from the electronics). Above of 200, however, the threshold excess rate falls off much less sharply, and threshold excess counts continue in a broad tail to very high thresholds. When running the cosmic ray trigger, I typically set the threshold near the start of this broad tail. Most of the variation between antennas is due to the discrete attenuation settings not being set to perfectly equalize the signal power at the ADCs. The notable outlier (green points in Figure 4.3) may have an incorrect attenuation setting in place.

Note that since the thresholds are set consecutively, each counts over a different 1.3-second interval. If the thresholds could be measured simultaneously, Figure 4.3 would be the complement of the cumulative distribution of the the samples. If the time series were a stationary stochastic process—as would be the case for pure Gaussian-distributed white noise—then measuring the thresholds one at a time would approach the complement of the cumulative distribution of the samples in any single 1.3-second counting window. However, the time series is not stationary, due to a constantly changing RFI background. Thus, the threshold scan results are not necessarily monotonic (see section 4.5 for an extreme case).

Trigger threshold scans

Figure 4.4 shows the results of a core-trigger-rate threshold scan. This test counts the number of times the trigger condition is met (threshold number of antennas

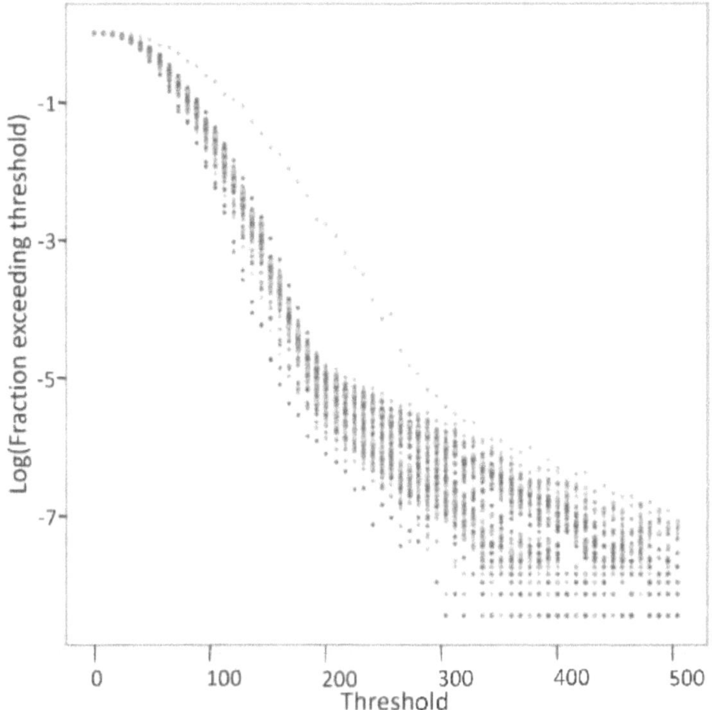

Figure 4.3: Threshold scan results from July 4 2023, for antennas from one representative SNAP2 board as an example. Results from each of the 64 single-dipole signals processed by the SNAP2 board are plotted in a different color. For each antenna and for each voltage threshold [in ADC units] on the x-axis, the SNAP2 counted how many clock cycles the filtered, smoothed power timeseries exceeded the square of the threshold, during a fixed number of clock cycles corresponding to 1.3 seconds. Note that while the threshold displayed is measured in ADC units for ease of comparing to the scale of the ADC range, the threshold set in the firmware is the square of the threshold shown here (a power threshold). Due to the FIR filter and power smoothing before the threshold block, saturation of the x-axis scale in this plot does not necessarily imply saturation of the ADCs. The y axis shows the base-ten logarithm of the fraction of clock cycles within the counting window that had a threshold excess. All 64 timeseries are thresholded simultaneously, but the different thresholds are set successively.

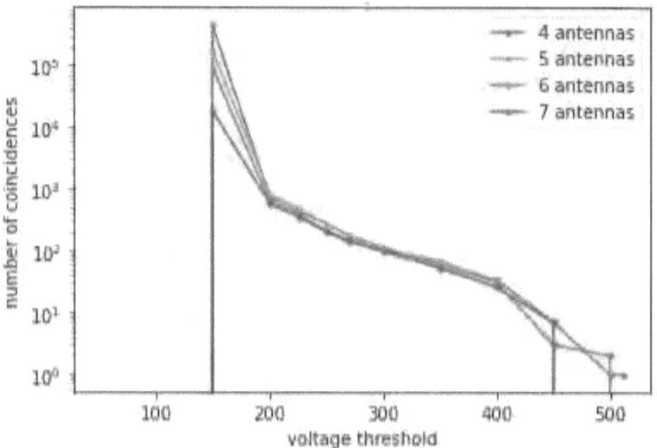

Figure 4.4: Core-trigger threshold scan for a single SNAP2 board. The x axis is the square root of the power threshold sent to the firmware. It is in ADC units. The y axis is the number of times the trigger condition was met, during the 1.3-second counting window. Different colors indicate different repetitions of the scan, for a different threshold number of antennas.

exceeding the threshold within the light travel time across the core), for different choices of power threshold and threshold number of antennas. No veto is running in this test. Similar to Figure 4.3, the results are plotted agains the square root of the power threshold. The measurements are not simultaneous, such that each point on the plot is measured by counting over a separate 1.3-second interval. Similar to the individual-antenna threshold scan, the trigger rate falls off more gradually for thresholds above 200. Counterintuitively, the number of triggers drops to zero at low thresholds. This feature is due to the fact that since the trigger condition counts the number of antennas whose smoothed power timeseries exceed the threshold within the light travel time across the core, if antennas cross the threshold repeatedly within that light travel time, the trigger condition is met for a longer amount of time. Multiple events can blend together and not be counted individually. For very low thresholds each antenna crosses the threshold so frequently that discrete instances of meeting the trigger condition do not exist.

Rate logging

Threshold scans run on the order of minutes. Another useful diagnostic of system performance and RFI environment has been logging statistics such as trigger and veto rates overnight. As an example, this section presents the rate of triggers, vetos, and other useful diagnostics logged overnight on July 4 2023. Due to limited disk space that night, the event snapshots read out from the buffer were saved during the first two hours, but event rate logging continued for a total of eight hours. The trigger and veto ran with the following settings:

- Power threshold for trigger antennas= 90000

- Power threshold for veto antennas= 90000

- Trigger window = number of clock cycles to travel 100 meters

- Veto window = number of clock cycles to travel 2000 meters

- Threshold number of core antennas = 6

- Threshold number of veto antennas = 1

The rate logging began at approximately 10:30pm local time. Figure 4.5 shows the rate of core triggers, vetos, and successful (un-vetoed) triggers measured over a 1.3-second interval every 30 seconds for 8 hours on the night of July 4-5 2023. These rates all experience an intense, brief increase once per hour. These hourly spikes are due to RFI from transmissions from water-level monitoring sensors operated by the Los Angeles Department of Water and Power, which controls much of the Owens Valley although 200 miles north of Los Angeles. This RFI appears in a single rate measurement each hour and therefore must be less than 30 seconds in duration. For the rest of this night, the veto is blocking more triggers in the first board shown than the second.

Figure 4.6 shows the rate that buffer readouts completed, with results from each SNAP2 board that was running the trigger overlaid in a different color. The SNAP2 boards are zero-indexed in this plot, and boards 0 and 1 are not shown. They did not run the trigger because they process antennas from the sparse array at middle distances and do not contain core-array antennas. These boards did send their data when the other SNAP2 boards triggered. The readout rate remains roughly steady for most of the night, punctuated by the hourly spikes in event rate. Since the system is designed for all SNAP2 boards to readout their data whenever any SNAP2 board triggers, the readout rate should be the same for all the boards. The fact that it is not identical during some of the intense hourly RFI events remains to be investigated but may be due to not reading the values from each board exactly simultaneously. The readouts from the intense hourly bursts of RFI do not result in similarly intense bursts of data recording, because the bursts exceed the maximum packet rate that can be received, and the excess packets are simply dropped either by the switch or the receiving computer. It is unlikely that a cosmic ray would be identifiable amid this RFI, and the episode is less than 30 seconds per hour; thus dropping the data is a convenient response.

Figure 4.7 shows the ratio of buffer readouts to successful triggers, with a successful trigger being a core-trigger that is not vetoed. Since the core trigger rate is specific to each SNAP2 whereas the readout rate depends on trigger signals communicated among the boards, there is not a one to one ratio between successful triggers and readouts. There tend to be more readouts than triggers since the readout rate reflects the triggers occurring in the entire array. Occasionally the ratio drops below one, because core triggers can continue to be counted while the board is in readout state. Thus, the intense hourly RFI shows up as increases in the ratio of readouts to successful triggers for some boards and as decreases for others.

Figure 4.8 shows the duty cycle of the veto, i.e. the fraction of clock cycles that the veto logic was blocking. Averaged over the entire eight hours, including the RFI bursts, the veto duty cycle is 0.1% for the SNAP2 board with the highest average, and lower for the others. During the intense RFI bursts, the veto duty cycle becomes large, as it should since there is not good data to be had during that time. This plot suggests that overall, running the veto contributes less than 0.1% to the total detector dead time.

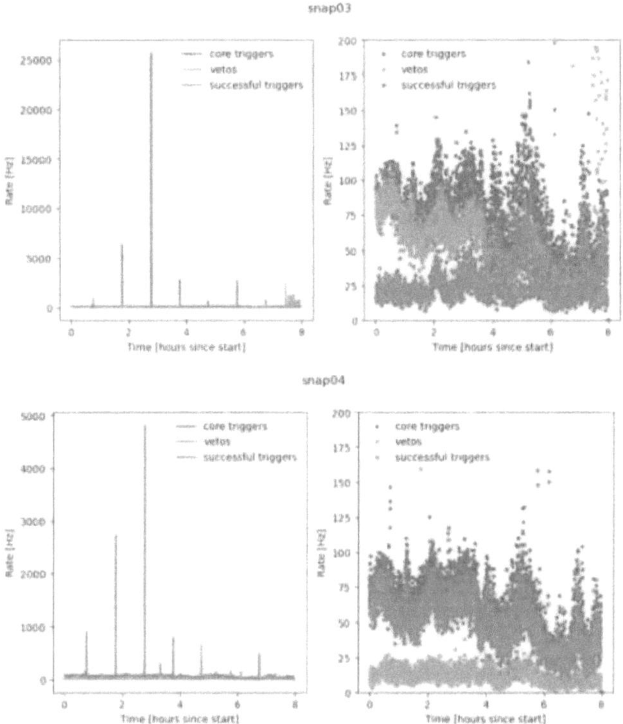

Figure 4.5: Rate of core triggers, vetos, and successful (un-vetoed) triggers measured over a 1.3-second interval every 30 seconds for 8 hours on the night of July 4-5 2023. The top and bottom pair of plots show the rates measured by SNAP2 #3 and SNAP2 #4, respectively (with board number 1-indexed). The difference between the left and right plots is the extent of the y axis.

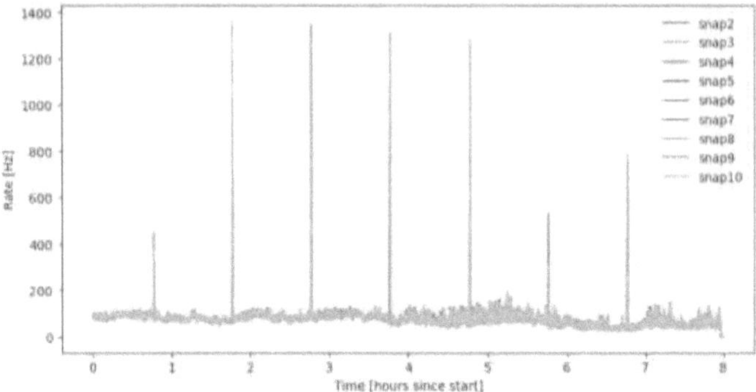

Figure 4.6: Rate that buffer readouts completed, measured over a 1.3-second interval once per 30 seconds over eight hours. Results for all the SNAP2 boards that were running the trigger are overlaid.

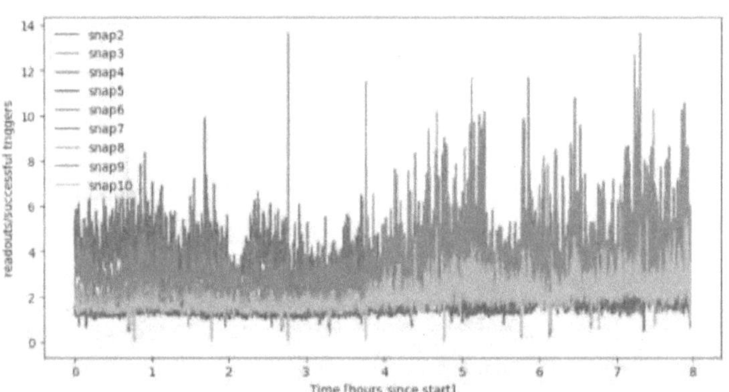

Figure 4.7: Ratio of buffer readouts to successful triggers, measured over a 1.3-second interval once per 30 seconds over eight hours. Results for all the SNAP2 boards that were running the trigger are overlaid in different colors.

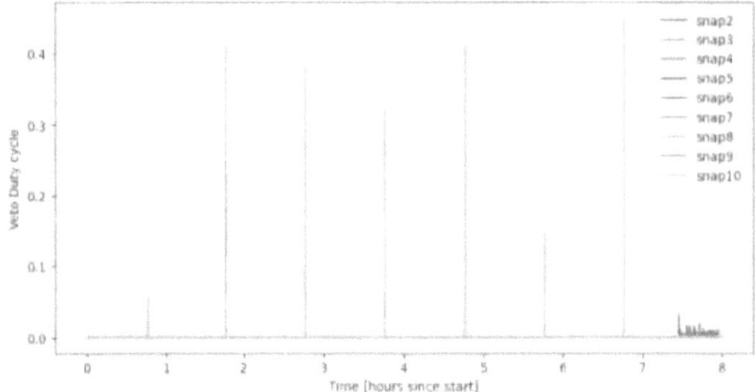

Figure 4.8: Veto duty cycle, measured over a 1.3-second interval once per 30 seconds over eight hours. The veto duty cycle is the fraction of clock cycles during the counting interval that the veto antennas were vetoing. Results for all the SNAP2 boards that were running the trigger are overlaid in different colors.

4.5 Defining a Signal to noise ratio and choosing to add a bandpass filter to the digital signal processing

Before this chapter moves from event rate statistics to event analysis, some discussion of the meaning of a signal-to-noise ratio is required. Signal-to-noise ratios (SNR) are a useful summary statistic to compare the significance of different events, but many different definitions appear in the cosmic ray air shower literature (see [89] for a summary). The reason so many definitions exist is that 'signal' and 'noise' are not fundamental physical quantities. They tend to be defined in terms of different physical quantities in different situations. The best definition of the signal-to-noise ratio depends on how that SNR is intended to be used, as well as on the nature of the signal and noise in question.

Cosmic ray radio emission must be identified against a background of non-cosmic ray radio-frequency power. This background consists of (1) thermal noise, mostly due to Galactic synchrotron emission, (2) narrow-band radio frequency interference from human activities, and (3) non-cosmic ray impulsive transients. Ideally, thermal noise would limit the background, but narrow-band RFI can affect the noise floor. At the individual-antenna level, although non-cosmic ray impulsive transients are a background to the cosmic ray emission of interest, their brevity means that cosmic ray emission will not need to be compared to a simultaneous impulsive RFI event. In other words, the impulsive RFI transients are separate events whose signal-to-noise ratio can be computed with the same definition that I adopt for individual-antenna cosmic ray signals.

This section discusses several closely linked design and analysis decisions. Choosing which metric to use for defining a signal-to-noise ratio led to the decision to smooth the power timeseries with a four-point average before the thresholding in the firmware (as well as software analysis). The decision to include a bandpass filter in the firmware and software signal processing was part of the same goal of reducing the random variability in the SNR for different antennas within an event, as well as stabilizing the trigger rate in the presence of RFI.

Defining a signal to noise ratio

Here, I wish to define an individual-antenna SNR—that is, choose some metric for the signal-to-noise ratio of the recorded timeseries of a single dipole for a single event. By event, I mean the change in electric field that triggered the cosmic ray buffer to read out. The main contributors to the electric field at the antenna are the impulsive transient event (if present), Galactic synchrotron emission, and narrow band RFI that is persistent (or at least persistent on longer timescales than the fast transients to be searched for). Non-diffuse astronomical sources such as extragalactic AGN, Galactic supernova remnants, and Jupiter make a smaller contribution to the total electric field. Since these commissioning observations take place at night time, solar radio emission does not contribute. The occasional presence of other sources such as lightning has not yet been explored.

The time series of voltages sampled by the ADC contains a component that is proportional to the electric field described above, as well as random voltage fluctuations contributed by the electronics along the signal path. The signals of interest to me are the impulsive transients, and so my definition of signal must be a quantity that varies with that contribution to the time series. My definition of noise for the purpose of this simple signal to noise metric will encompass all the other background components. Note that this estimate of noise power will include components that in other contexts would be considered a signal. There are signals from various astronomical sources that are interesting in their own right in other contexts. The 27 MHz narrow-band RFI was transmitted by someone for some purpose, and to a CB radio enthusiast (see section 4.3) this component would be the signal. Since the signal to noise ratio is used to assess significance, the definition depends on what is significant in the context.

My purpose for defining an SNR is to have a quick way to check which antennas detected an event and assess the significance of the detection, by checking which antennas experienced a change in the electric field that was not explicable by the noise, and specifically how much stronger than the noise it was. Given this purpose of the SNR statistic, it remains to identify a good definition of the signal and the noise.

A key choice is whether to use a ratio of powers or a ratio of amplitudes (i.e. voltage or electric field, see e.g. [89] for more discussion). In earlier tests I used an amplitude-like definition, defining the signal-to-noise ratio as the ratio of the maximum ADC sample in the buffer to the standard deviation of the ADC samples before the event. At the time, the firmware applied the power threshold to the simple square of the ADC voltages and thus was equivalent to thresholding on the absolute value of the voltage timeseries, making an amplitude SNR a natural choice.

The main disadvantage of this approach is that the peak electric field from a cosmic ray air shower may arrive very close to an instant that is sampled by the ADC, or may arrive between ADC samples. After passing through the analog electronics and digital filter, the resulting signal may have a single large peak on one sample or smaller peaks on two adjacent samples, for example. This sensitivity of the peak value to the phase of the clock cycle could contribute to some large differences in SNR observed between nearby antennas. To make an amplitude definition of the signal that solves this problem, one could take the peak electric field of the magnitude of the analytic signal. Obtaining this value involves a Hilbert transform, and it is more useful here to define the signal in a way that is simple to implement in the FPGA firmware.

The impulse response of the digital FIR spreads the power of an impulsive transient over several samples. Even without the digital FIR, the impulse response of the analog electronics (with a filter of similar bandwidth) spreads the power similarly. Averaging the power over several consecutive samples is not equivalent to finding the total power under the Hilbert envelope, but gives a simple approximation that is easily implemented on the FPGAs.

Thus, I define the signal as the peak value of the power timeseries after smoothing the power timeseries by a four-sample moving window average. The window length of four samples is chosen to be close to the impulse response time, thereby integrating the power over the duration of an impulsive transient. I define the noise as the mean of the power timeseries. The thresholding block in the firmware trigger operates on the filtered, smoothed power time series.

The following equations summarize this processing for the time series of a single dipole, including the filtering step which will be detailed in the next subsection. Note that the threshold in the FPGA firmware operates on the last of the timeseries.

$$v_f[k] = (h_1 * v)[k] \tag{4.1}$$

$$p[k] = (v_f[k])^2 \tag{4.2}$$

$$p_s[k] = (h_2 * p)[k] \tag{4.3}$$

In the above, $v[k]$ is the ADC voltage on time sample k in the saved buffer output (after the transmitted data has been truncated to remove the constant number of unusable time samples at the edges of the buffer), $v_f[k]$ is the filtered voltage timeseries, $p_s[k]$ is the power time series obtained from the square of $v_f[k]$, and $p_s[k]$ is the smoothed power time series. The FIR filter coefficients are denoted by h_1 and the four-sample moving window average is accomplished by setting $h_2 = [0.25, 0.25, 0.25, 0.25]$. The symbol $*$ denotes the convolution, which is performed with the Python function scipy.signal.convolve [96], using the option to truncate the output to valid data. This truncation means the time indices of the filtered and unfiltered time series are not directly comparable; if $v[k]$ were a delta function with peak at sample k_1, the peak of the impulse response in $v_f[k]$ occurs at $k_2 < k_1$, where $k_1 - k_2$ is a constant determined by the length of the filter.

Note that to relate this SNR to the probability of a fluctuation occurring by random chance in a thermal noise distribution, the noise estimate could be decreased by a factor of two to account for the power smoothing (with the uncertainty on the mean of N samples going as $\frac{1}{\sqrt{N}}$. The time series is not pure thermal noise, however, and in section 4.6 I estimate the significance of events by comparing to the empirically-measured median SNR computed for a snapshot of pure noise. In general, background signals such as narrow-band RFI should be kept distinct from fundamental noise (such as from thermal processes) in the system, and in the next section I introduce a bandpass filter to help suppress the non-impulsive RFI background.

Decision to include a digital bandpass filter

This section discusses the decision to add a digital finite impulse response (FIR) bandpass filter before the power smoothing step, in the firmware and the software signal processing. Background non-impulsive RFI changes the noise floor and can contribute to whether or not an antenna's signal crosses the threshold. The strongest non-impulsive, narrow-band RFI occurs below 20 MHz, in the FM band (above 85 MHz) and at times in the CB band at 27 MHz. Even with the suppression provided by the analog filter, the RFI below 20 MHz and in the FM band can make a non-negligible contribution to the total power, and the the 27 MHz RFI is in the pass band of the 18–85 MHz analog filter. Furthermore, if the electric field at an antenna consists of the superposition of (1) thermal noise, (2) an impulse arriving from one direction, and (3) a tone arriving from a different direction, then the measured peak of the impulse will depend on the phase of the tone at the antenna. The large power in narrow-band RFI motivated adding a digital bandpass filter in the firmware and in the software processing.

The impulse response and magnitude response of the filter currently in use are shown in Figure 4.9. This filter is a 24th order bandpass finite impulse response filter. Designing the filter involved two different tradeoffs: (1) peak of the impulse response vs width of the passband, (2) sharpness of the pass band cutoffs vs sharpness of impulse response, and (3) length of the impulse response vs shape of the pass band and strength of the out of band suppression.

A finite impulse response filter of length N takes an input timeseries and outputs a timeseries that consists of the input convolved with the series of N filter coefficients. The output of the filter on sample k depends only on the input samples from k-N to k, and hence the impulse response is finite. The impulse response is defined as the output timeseries for an impulse input[2]. For an FIR, the impulse response simply equals the series of filter coefficients. The shape of the frequency response is the Fourier transform of the impulse response.

Cosmic ray signals are impulsive and broadband. Thus, the narrower the passband of the filter, the more power from the cosmic ray is lost, but the substantial improvement to the noise floor by suppressing the RFI can make this loss worthwhile. Suppressing the narrowband RFI requires a passband at most from 28 to 85 MHz, thus roughly half the baseband will be suppressed and so the peak height of the impulse response

---

[2]An impulse here is a timeseries that equals 1 for k=0 and equals zero for all $k \neq 0$. It is the discrete-time analogy of a delta function.

is ~ 0.5. This FIR impulse response however does not mean that half the power in a cosmic ray signal is lost by introducing this FIR, since by the time the cosmic ray signal reaches the digital FIR it has already passed through an analog FIR of a similar bandwidth. The digital FIR provides additional suppression at frequencies that have already been suppressed by the analog electronics.

An ideal frequency response might be perfectly flat across the passband, zero elsewhere, and would maximize the bandwidth of the passband by pushing the cutoff frequencies very close to the edge of the frequencies affected by the RFI. Such a bandpass response is impossible to achieve with a finite number of filter coefficients, and furthermore would have excessive ringing in the impulse response, because the Fourier transform of a boxcar window is a sinc function. The more gradual the roll-off from passband to stopband, the less ringing in the impulse response. For this reason, and in order to limit the number of coefficients required, the pass band is chosen to be 35-80 MHz which allows a gradual roll-off to large suppression at the frequencies of the RFI. The 35 and 80 MHz boundaries of the pass band are the frequencies by which the magnitude response has fallen to 3 dB (i.e. roughly half power). The magnitude response is roughly flat across the passband but not flat in the stop band, with several nulls. I adjusted the lower edge of the passband to place one of these nulls over the 27 MHz RFI.

The longer the filter, the more resource-intensive it is to implement. The filter currently in use is a 24th order filter (i.e. there are 25 coefficients). I chose this size because it was the smallest which allows the impulse response to ring down to zero at the ends. The current implementation in the firmware uses roughly as many digital-signal-processing hardware (DSPs) components as the entire rest of the OVRO-LWA firmware. Some alternate strategies to use fewer resources will be pursued in the future, including for example implementing a bandpass filter by digital down-conversion followed by a low-pass FIR. Infinite impulse response filters can achieve comparable frequency responses to FIRs with much fewer coefficients, but are not considered a good option here since this is a search for impulsive transients (hence an infinite impulse response would create difficulties).

The FIR was added to the firmware at the same time as deciding to introduce the four-sample smoothing before the threshold block in the firmware, and in the SNR definition used in the software processing.

Results

In summary of the results presented in this section, adding the FIR stabilized the trigger rate. Figure 4.10 shows individual-antenna threshold scans for several different days, before and after changing the firmware from applying a threshold on the square of the raw voltage timeseries to applying the threshold after filtering the voltages and smoothing the power. For each plot, the trigger and veto threshold blocks were used to make two simultaneous threshold excess measurements for the same antenna; the trigger and veto subsystem was not running a cosmic ray search at the time. During the scan, the values of the core threshold started from zero and increased on each step while the values of the veto threshold decreased from 512 on each step. In this way, after the whole scan, there are two measurements from different times at each threshold.

The upper row of plots are from January 2023. The upper left shows a scan from an exceptionally RFI-quiet time. The tail at high thresholds is barely present. The upper right plot shows more significant RFI. The difference between the blue and red points gives a sense of the time variability. The middle left plot, also from January 2023, shows a time of exceptionally bad RFI. In this plot, the threshold-excess rate appears to vary more strongly with time than with threshold and no threshold could be set to meaningfully run the trigger and veto system. The middle right plot shows an example from April 2023.

Although the RFI conditions vary between days, as illustrated by the different examples from January 2023, threshold scans as unstable as the middle left plot of Figure 4.10 have never been observed since adding the FIR and power smoothing. The bottom row shows two threshold scans after the filter and power smoothing logic was compiled in the firmware (left June 30 2023, right July 4 2023). There is much less difference between the blue and red points, and the points follow a smoother curve. Thus, this change appears to have made the threshold excess rates more stable. One further test that would improve this result would be to take new data that uses the same number of threshold steps as the early plots to more directly compare. The plots in January take more steps, which means that the whole scan takes longer, allowing more time for the RFI to vary. This time difference is not likely to explain the entire improvement in the smoothness of the scans. It is not possible to rapidly change between firmware versions or to perform a threshold scans with and without the filter simultaneously for the same antennas, but another interesting comparison test would be to one version of the firmware on one SNAP2 board and another on a

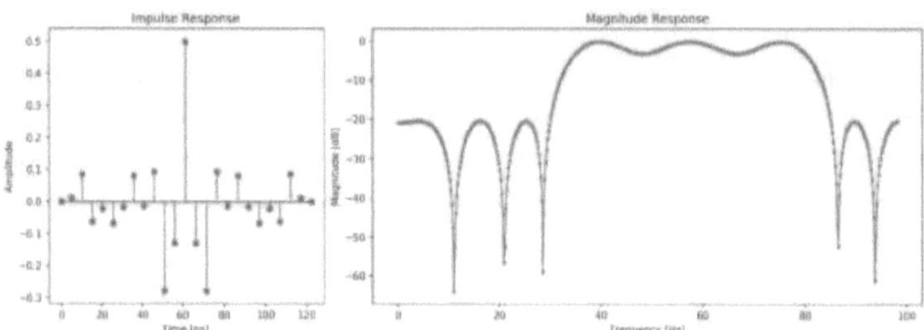

Figure 4.9: Left: Impulse response of the 24th order FIR bandpass filter used in the firmware trigger and the software processing of recorded timeseries. Right: Magnitude response of the FIR filter. The passband is 35–80 MHz.

different SNAP2 board, to make a comparison of threshold scans during the same RFI environment. The threshold scans described above, however, strongly suggest that the FIR and power smoothing firmware have stabilized the trigger rate.

Figure 4.11 shows the SNR (as defined above) as a function of antenna position for an event from the night of July 1-2 2023, plotted with and without the filter applied. The trigger was running with the FIR and power smoothing in place, but since the recorded data is always the raw voltages it is possible to choose in software whether or not to apply the filter. In the top row, the filter is not applied, and in the lower row, the filter is applied. For this event, applying the filter improves the signal to noise ratio. This particular event is also discussed further in section 4.6. Note that a few antennas appear in one panel but not the other, because antennas that do not meet the mean power and kurtosis criteria described in 4.6 are excluded from the plot. These criteria were introduced to automate the identification of antennas are not operational, but strong RFI can cause antennas to be excluded from this cut. The most readily notable example in Figure 4.11 is the antenna near -500m, -750m, which only passes the signal quality cuts after applying the bandpass filter.

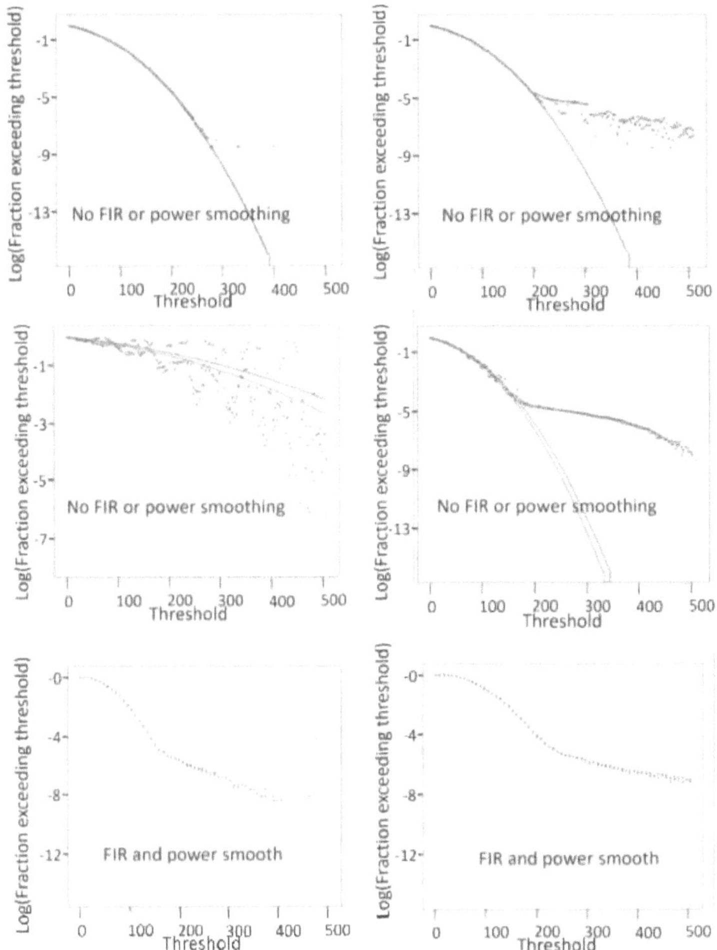

Figure 4.10: Single-antenna threshold scans in different conditions, before and after compiling the FIR in the firmware. For each plot, the core trigger (red points) and veto (blue points) threshold blocks were used to make two simultaneous threshold excess measurements for the same antenna; the trigger and veto subsystem was not running a cosmic ray search at the time. The threshold axis shows the square root of the threshold applied in firmware, in order to present a voltage value in ADC units. See the text for a complete description of each panel.

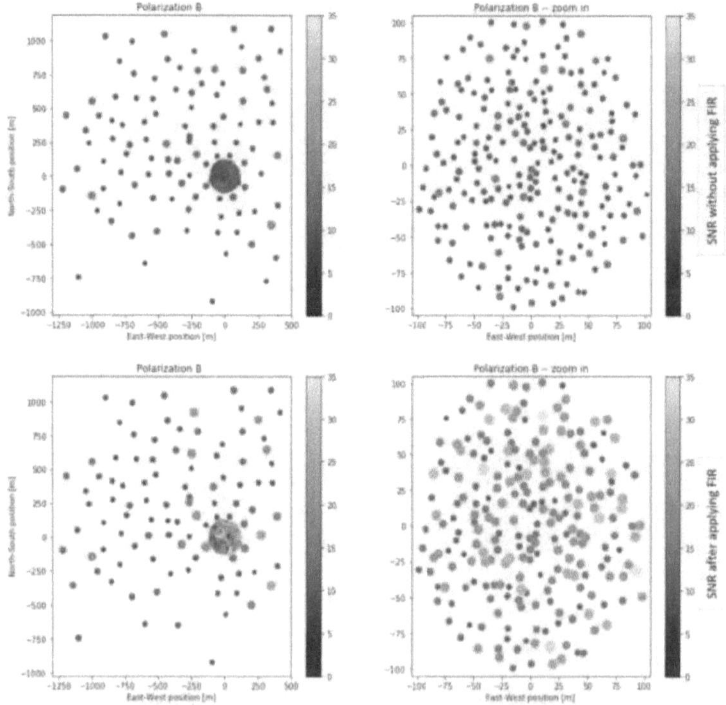

Figure 4.11: Signal-to-noise ratio calculated without (top row) and after (lower row) applying the bandpass FIR filter. The color and spot size correspond to the SNR, displayed for each E-W polarized dipole overlaid on antenna position. In each row, the left panel shows the whole array and the right panel zooms in on the core array. Only the E-W polarization is shown for this polarized event.

4.6 Defining Criteria for Preliminary Cuts

This section discusses the development of the first stage of RFI rejection in the software processing. The final software processing to identify cosmic rays will involve comparing the cosmic ray signals to the lateral power distribution and polarization expected for cosmic ray emission (see Figure 4.1). Some intermediate cuts may also be introduced in the future, such as using the arrival direction to efficiently reject RFI from known sources.

The purpose of the preliminary cuts is to reject as much RFI as possible by the simplest metrics possible, to reduce the number of events that require detailed processing, while limiting the chance of inadvertently rejecting a cosmic ray.

Within the preliminary cuts, the overall steps are summarized as follows:

1. Load the saved event data.

2. Calculate individual-antenna statistics to determine the presence of (1) a likely-impulsive signal, (2) significant steady RFI, (3) a thermal background, or (4) no meaningful signal (e.g. disconnected antenna).

3. Exclude faulty antennas. In practice, this cut serendipitously excludes certain types of RFI contamination as well.

4. Calculate event-based summary statistics by combining the individual-antenna statistics.

5. Apply cuts to events using the event-based summary statistics, to reject RFI events and identify events worth further inspection.

Obtaining smoothed, filtered power timeseries

Event data is saved to disk in the order that packets arrive. Each packet has a header with metadata including information to identify the dipole the data corresponds to as well as a time stamp and a summary of the settings for the trigger and veto system at the time the data was recorded (see Chapter 3) for complete details. The event classification software begins by loading saved data and using the timestamps to separate the series of packet data into discrete events. Packets from one FPGA all have the same timestamp, and packets from different FPGAs have timestamps that differ by the amount of time it took for each different FPGA to receive the trigger signal. In software processing, packet data is organized into a series of event snapshots by collecting all the packets with timestamps that differ by less than a maximum offset. This maximum offset is typically set as small as the time it takes for the trigger to travel around the loop of SNAP2 boards (137 clock cycles). This system of organizing the packets into events works because the timestamp difference between different FPGAs is much smaller than the readout dead time.

After loading data and organizing it into events, the first steps of the signal processing software replicates the first stages of processing that occur in the trigger firmware in the FPGA, following the equations in section 4.5. First, the finite impulse response bandpass filter is applied to the raw ADC voltage time series for each antenna. Then, this time series is squared to obtain a power timeseries, which is then smoothed with a four tap moving window average.

Single-dipole summary statistics

The following summary quantities are then calculated from the above timeseries:

1. Peak value of smoothed power timeseries: $p_{\text{peak}} = \text{maximum}(p_s)$, where p_s is defined as in equation 3 in section 4.5.

2. Sample index of peak value of smoothed power timeseries: k_{peak} is defined as the minimum (i.e. earliest) value of k for which $p_s[k] = p_{\text{peak}}$.

3. Mean smoothed power well before the trigger:

$$< p_{\text{s-before}} > = \frac{1}{2000} \sum_{k=0}^{k=1999} p_s[k]$$

4. Mean smoothed power soon after the trigger:

$$< p_{\text{s-after}} > = \frac{1}{k = k_{\text{peak+10}}} \sum_{k=0}^{k=k_{\text{peak+10}}} p_s[k]$$

5. Signal-to-noise ratio: $\text{SNR} = \frac{p_{\text{peak}}}{<p_{\text{s-before}}>}$

6. Excess kurtosis of the filtered voltage data before the trigger: $\kappa = \frac{\mu_4}{\sigma^4} - 3$ where μ_4 and σ are the fourth central moment and the standard deviation, respectively, of the first 2000 samples of v_f.

7. Change in mean power: $s = \frac{<p_{\text{s-after}}>}{<p_{\text{s-before}}>}$

Note that all the above statistics are calculated from the smoothed power time series, except for the excess kurtosis which is computed on the filtered voltage timeseries.

The current version of the code computes a few additional statistics for testing purposes, which are not detailed here as they are not used in this set of preliminary cuts.

Excluding faulty antennas

Some of the above statistics are used as metrics for assessing which antennas have usable signals. At present, only a few antennas have faults along their signal paths, but I developed this code during the time that new antennas were continuing to be added to the array towards the end of the hardware upgrade. I select all the antennas whose mean power and kurtosis before the trigger fall within an acceptable range. These ranges are typically $625 < (< p_{s-before} >) < 2500$ and $-1 < \kappa < 1$, but the appropriate mean power range in particular can change as the attenuation settings in the analog receiver boards continues to be optimized. The initial flagging of bad antennas relied entirely on the mean power, which is nearly always an effective way to identify antennas that have been disconnected, have broken components, or have an inappropriate attenuation setting applied. I added the kurtosis to the cut because one antenna with a fault along its signal path had an ADC output near zero for most samples except for occasional sudden ADC saturations, which occurred just often enough that the mean power fell within the typical range for a good antenna. Although the kurtosis cut was introduced to exclude faulty antennas, certain types of RFI are also rejected by checking the kurtosis.

Additionally, signals for which the peak of the mean smoothed power occurs near the end of the buffer are excluded by requiring $k_{peak} < 3944$, since for these events the change in power s_p is not defined. Impulsive transients that trigger the buffer to read out appear roughly three-quarters of the way through the buffer, and so a peak appearing in the last few samples of the buffer is not what triggered the data readout. Timeseries with peaks near the end of the buffer either did not detect the event above the background—the peak value of a timeseries of pure noise is equally likely to occur anywhere in the buffer—or, the event was not impulsive and consisted of a series of peaks.

Summary statistics for each event

After excluding antennas that do not meet the kurtosis and mean power criteria, summary statistics for each event are calculated from the set of summary statistics for individual dipoles and used to categorize events. For each event, I compute the number of dipoles with an SNR above a minimum cutoff, the number of veto antennas with peak above the veto threshold, the median power change ratio, and several metrics comparing the power in the core array to the outlying antennas.

For the count of the number of antennas above a minimum SNR, the cutoff value SNR>5 used in the analysis presented in this section was chosen to match the analysis in [65], but is less than the typical SNR of pure background, making this count not particularly useful. A higher threshold will be used in future versions of the cuts. The remainder of the cuts in this section differ from the analysis in [65].

The packet header data indicates whether the packet data corresponds to a dipole that was operating as a veto antenna at the time the triggered search was run, as well as including the values of the veto thresholds (power threshold and threshold number of antennas). Since the veto is initially applied within each FPGA, no more than the threshold number of veto antennas processed by one FPGA have a peak smoothed power above the veto power threshold in the saved data. In software, a more restrictive cut can be applied, with the criterion that no more than one of the veto antennas in use at the time exceeds the power threshold.

I compute several metrics to compare the power in the core antennas to the outlying antennas. For these metrics, the core antennas are defined as antennas within 150 meters of the array center, and the distant antennas are defined as those more than 250 meters of the array center (including many antennas which are not far enough to be used for the veto). The code separates the antennas into four groups: N-S polarization dipoles in the core, E-W polarization dipoles in the core, distant N-S polarization dipoles, and distant E-W polarization dipoles. For each group, the dipoles are ranked by SNR, and the following quantities are computed: maximum SNR, sum of the five largest SNRs, and sum of the ten largest SNRs.

Most events are polarized, and thus it is useful for most of the quick classification cuts to consider the data corresponding to the dominant polarization. Using the core antennas, the sums of the ten largest SNRs for each polarization are compared and the polarization with the larger sum is labelled as the dominant polarization of the event.

For the dominant polarization, two ratios are computed that compare the power in the core vs the distant antennas: the strongest core-antenna SNR divided by the strongest distant-antenna SNR, and the sum of the five strongest core-antenna SNRs divided by the sum of the top five strongest distant antenna SNRs. The preliminary cuts presented here include the criteria that these ratios be greater than 2.5. The

Cut	Number passing	Fraction of total
(0) complete events of 704 packets	527940	1
(1) Power before vs after $0.5 < s < 1.7$	39841	0.075
(2) ≤ 1 veto antenna above veto threshold	469523	0.889
(3) (max SNR in core)/(max SNR in distant)	445	8.43e-4
(4) (sum top 5 core SNRs)/(sum top 5 distant SNRs)	125	2.37e-4
Combination	50	9.47e-5

Table 4.1: Numbers of events from three hours of data collection on July 1-2 2023 that pass different cuts. Combination refers to passing cut (1), (4), and also having more than 50 dipoles (of either polarization) with SNR>5. The latter is not included as a row in the table because it was not significantly restrictive. Cuts (1), (3), and (4) are applied based on the dominant polarization.

cut is set at 2.5 by examining the distribution of these ratios for a dataset of 1000 events and choosing the value that separates the central core of the distribution from a handful of outliers.

Finally, a cut is applied on the ratio of the mean smoothed power before and after the event. For each event, I compute the median of s over all antennas that pass the individual-antenna cuts and the minimum SNR cut (although the latter is currently not restrictive), for each polarization separately. Events passing this cut must have $0.5 < s < 1.7$ for the dominant polarization. This cut helps reject non-impulsive RFI. Sometimes the trigger condition is met when a strong RFI episode begins. For these events, the power after the trigger is substantially higher than the power before. The 'after' time interval is chosen to start 10 samples after the peak in order to start after the time for the impulse response to ring down. I chose the cutoff values $0.5 < s < 1.7$ by examining the distribution of s for simulated timeseries of pure white noise: 97% of the simulated white noise time series have $0.5 < s < 1.7$.

Examining the July 1-2 2023 data set as an example

The preliminary cuts described above were used to explore a dataset of 527940 events from three hours of data collection on the night of July 1-2 2023. Table 4.1 summarizes the numbers of events that pass each of the cuts discussed above.

Figures 4.12 to 4.16 show data for example events that passed the following combination of the cuts described above:

1. The median s satisfied $0.5 < s < 1.7$.

2. More than 50 antennas had SNR > 5. (Not significantly restrictive, see discussion above.)

3. The sum of the five strongest core SNRs in the dominant polarization is more than 2.5 times the sum of the top five distant SNRs in that polarization.

Note that this subset of cuts does not include all the cuts described above, as different tests examine different subsets of cuts and a final version has not yet been finalized. The two events chosen illustrate two of the most typical types of transient but do not encompass every type of RFI observed.

Figure 4.12 shows an event with a tightly localized power distribution. The time series for this event (4.13) are clearly RFI. This event may be emitted by the electronics processing shelter itself, as it is brightest in the part of the array nearest to the shelter and falls off sharply with distance. Note that the ADCs are saturating for the example antenna shown. The event passed the cut on the change in power before and after the trigger because this cut uses the median power, and relatively few antennas have a strong detection.

Figure 4.14 shows an event with a more scattered power distribution that does tend to concentrate on the core but is not clearly cosmic ray like. Unlike the first example, the time series for this event show an isolated impulsive transient (4.15). Figure 4.16 shows the arrival time of the peak plotted over the antenna position. The arrival time is calculated as $k_{\text{peak}} + t_0 - t_{\text{min}}$, where t_0 is the packet timestamp and t_{min} is the minimum of all the packet timestamps. The packet timestamp t_0 must be added in order to be able to compare arrival times between different SNAP2 boards. Scaling from the minimum timestamp simply provides a convenient way to avoid handling 64 bit integer timestamps in the plotting code. From the plot of arrival times, this event appears to arrive from the south and sweep north, although there are a number of outliers among the antennas (potentially due to low signal to noise, but not all have been confirmed).

Antenna mis-alignments could cause scatter in the single-polarization SNR plots. During construction, antennas were roughly aligned with the North-South dipole aligned to geographic north, but as a final step in the array construction, all the antennas are undergoing a second alignment check. A few antennas were found to be loose enough to spin out of alignment, and a few were found to have been installed with a 90 degree rotation or have swapped cables between polarizations (which is equivalent to a 90 degree misalignment). I checked whether smaller misalignments

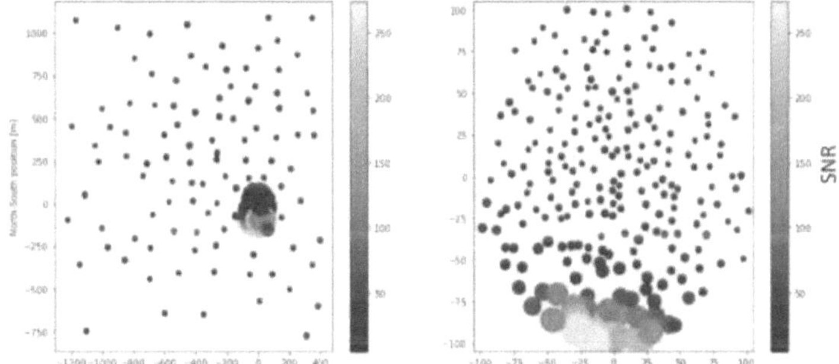

Figure 4.12: SNR vs antenna position for an example event from the night of July 1-2 2023. In both plots, color and spot size are proportional to the SNR, and spot location indicates antenna location. Only antennas with mean smoothed power between 900 and 2500 ADC units, and kurtosis between -1 and 1 are shown. Left: whole array. Right: zoomed in on core array. Only data from the N-S polarization is shown.

could cause the scatter in the signal-to-noise plots by plotting the mean of the signal to noise in the two polarizations. Although a few antennas that were low-SNR outliers in a single-polarization plot corresponded to high-SNR outliers in the other polarization, the overall scatter across the array was not substantially reduced. The misalignments appear to have affected only a very small fraction of antennas.

Injecting simulated events

This section describes a test using simple impulse signals injected in un-triggered snapshots of background timeseries in order to test the version of the preliminary cuts described above. I collected the background dataset with a script to read out

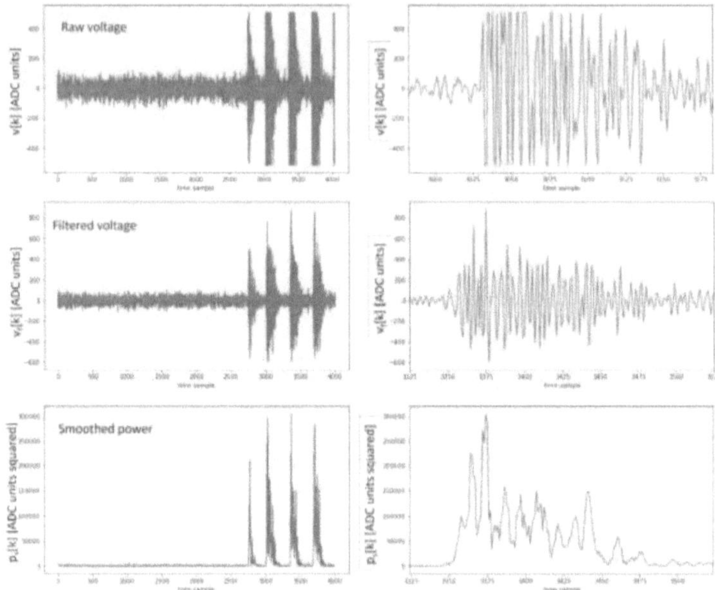

Figure 4.13: Timeseries for one antenna (the core antenna with the strongest N-S Polarization detection) shown in Figure 4.12. The left panel shows the entire timeseries and the right panel shows the timeseries zoomed in on 200 samples near the peak. Top: Raw ADC voltage output. Middle: Filtered voltages. Bottom: Smoothed power timeseries.

the buffer from a software trigger every 30 seconds from roughly 10:00PM PDT to 6:30AM on the night of 3-4 July 2023, for a total of 1000 snapshots. Using this dataset, the simulation injected an impulsive signal onto the timeseries from one dipole for ten core antennas. Injecting the impulsive signal consisted of taking the sample 2500 in the timeseries and adding a fixed amplitude to it. If the result was larger than 512, the value was set to 512. All 10 antennas received the same impulse. This injected signal did not take into account the impulse response of the analog signal path or the lateral distribution function of a real cosmic ray air shower, but for the purposes of testing the cuts described in this chapter it created a signal that was localized to the the core. Additionally, typical veto antennas were assigned veto roles, and a typical veto power threshold of 90000 was set. The resulting dataset was processed through the preliminary cuts described above, with the following

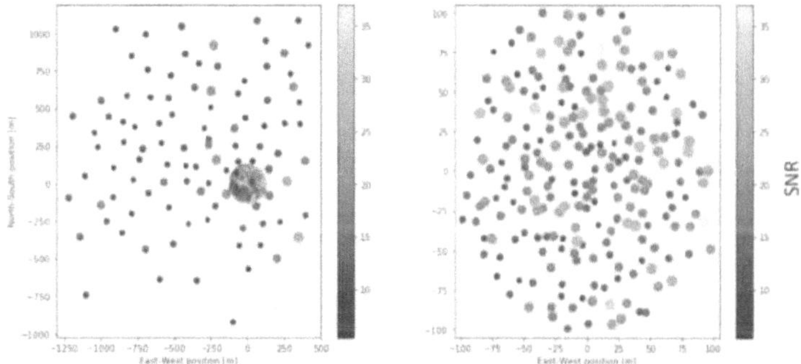

Figure 4.14: SNR vs antenna position for an example event from the night of July 1-2 2023. In both plots, color and spot size are proportional to the SNR, and spot location indicates antenna location. Only antennas with mean smoothed power between 900 and 2500 ADC units, and kurtosis between -1 and 1 are shown. Left: whole array. Upper right: zoomed in on core array. Only the data from the dominant polarization dipoles (E-W) are shown.

parameters:

Parameters for per-antenna cuts

1. maximum mean power: 2500

2. minimum mean power: 625

3. Minimum SNR: 5

4. Minimum excess kurtosis: -1

5. Maximum excess kurtosis: 1

Parameters for event-based cuts

1. Minimum number of antennas above minimum signal to noise ratio: 50

2. Median ratio of power after event to power before > 0.5

3. Median ratio of power after event to power before < 1.7

4. Ratio of maximum core SNR to maximum far SNR > 2.5

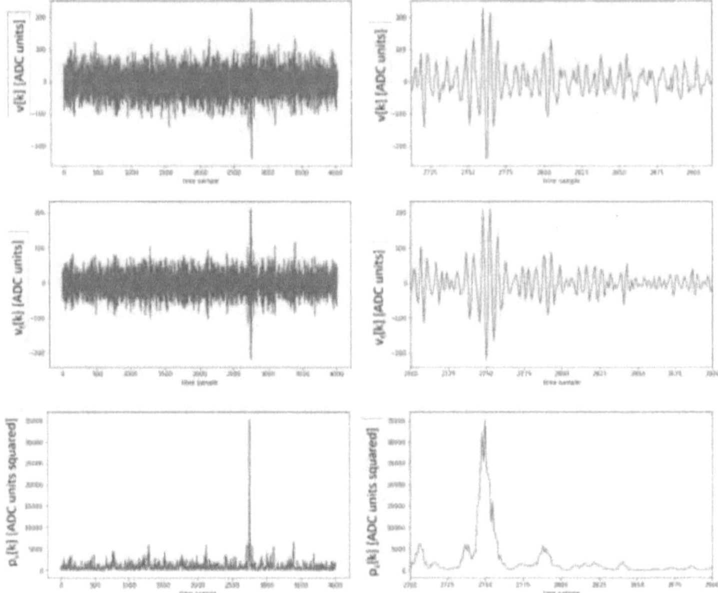

Figure 4.15: Timeseries for one antenna with a strong detection shown from Figure 4.14. In each row, the left panel shows the entire timeseries and the right panel shows the timeseries zoomed in on 200 samples near the peak. Top: Raw ADC voltage output. Middle: Filtered voltages. Bottom: Smoothed power timeseries.

5. Ratio of sum of top 5 core SNRs to sum of top 5 cor far SNRs > 2.5

6. Maximum number of veto detections: 1

The test was repeated for each polarization and for the following amplitudes of injected impulse: 500, 400, 300, 200, 0 (no pulse injected).

The fraction of simulated events passing the cut on the change in the power before and after the event and the cut on the number of veto detections the same for all injected pulse amplitudes, because the background data set was the same in each case and the injected signal did not affect the veto antennas or the samples in the window after the event. 99.6% of simulated events passed the cut on the change in power, and 99.8% of events passed the cut on the number of veto detections.

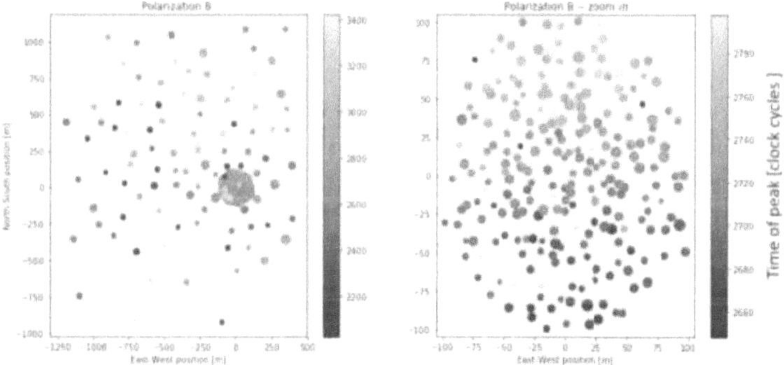

Figure 4.16: Arrival time of peak vs antenna position for the example event shown in Figure 4.14. In both plots, color is proportional to the peak arrival time (in clock cycles since first buffer timestamp of snapshot), the spot size is proportional to the SNR, and spot location indicates antenna location. Only antennas with mean smoothed power between 900 and 2500 ADC units, and kurtosis between -1 and 1 are shown. Left: whole array. Right: zoomed in on core array. Only data from the dominant polarization dipoles (E-W, labelled 'polarization B') are shown.

Based on the dataset without any pulse injected, the median SNR of pure background is 7.3. Figure 4.17 shows the median SNR among the antennas with a signal injected, as a function of the amplitude of the injected impulse. In this plot, two SNRs are shown. One SNR uses the maximum of the smoothed power timeseries (with the FIR applied to the voltages) in the entire buffer, and is denoted "total buffer" in the plot. This SNR is what the preliminary cuts code uses. The other SNR uses the maximum value in a small window around the sample corresponding to the known injected impulse. This SNR is denoted "window" in the plot, and describes the SNR of the injected pulse even if it is below the noise level of the buffer as a whole. At large pulse amplitudes the two SNRs are the same because the injected impulse creates the strongest peak in the timeseries. At low impulse amplitudes, the injected impulse is below the noise such that the peak value likely occurs elsewhere in the timeseries.

The number of events passing each of the cuts that depend on the ratio of SNRs in core and distant antennas are shown in Figure 4.18, plotted as a function of SNR. In this plot the median window SNR corresponding to the injected pulse amplitude is used, because this SNR better describes the injected impulse than taking the peak of the whole timeseries. For the strongest injected signals, 70% of events pass the cut using the maximum SNR, and less than half pass the cut using the sum of the top 5 SNRs. These results suggest that this version of this cut is too restrictive, and would exclude a substantial number of cosmic rays. The final version of the preliminary cuts should search for an excess of power in the core in a more robust way, or rely on additional metrics as well.

To begin updating the cuts, I increased the power SNR for a detection to 16 (approximately twice the median SNR of background snapshots). Additionally I relaxed the minimum ratio comparing the top SNR in the core to the distant antennas. In the version discussed above, I used a cutoff of 2.5. To find a better cut, I repeatedly ran an abbreviated simulation using 10% of the background snapshots from the simulations above, lowering the cutoff each time until 99% and 96% of simulated events with an amplitude of 500 made it through the cuts that use the ratio of the maximum SNRs and the ratio of the sum of the top five, respectively.

After processing a dataset of ∼ 600000 triggered snapshots from the cosmic ray buffer with this updated cut, more than 10000 events passed the set of preliminary cuts, which is three orders of magnitude more than before. In the future, another type of quick cut to search for power concentrated in the core array may be useful, but a reduction from 600000 events to 10000 events by the first stage cuts is sufficient for these passing events to progress to the proposed second stage of cuts involving searching for spatial and temporal clustering as indicators of specific RFI sources (see Figure 4.1).

4.7 Conclusion

In conclusion, the cosmic ray detector firmware successfully finds fast transients and reads out the buffer. Adding an FIR bandpass filter appears to have made the threshold excess rate more stable, and reduces the number of antennas that are rejected in software for having high kurtosis. For a definition of the signal to noise ratio as the maximum value of the filtered, smoothed, power timeseries divided by the mean power in this timeseries, the typical SNR of un-triggered background data is 7.3. The preliminary cuts described in this chapter are preliminary in two senses: they are the first version of the software tested with the intent to refine the

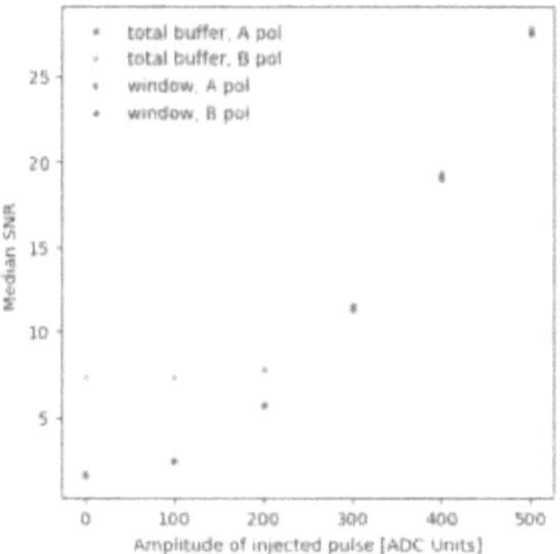

Figure 4.17: Median SNR among all dipoles with an impulse injected as a function of impulse amplitude, shown for two versions of SNR and both polarizations. In the legend, "total buffer" refers to the SNR calculated from the maximum of the smoothed power timeseries (with the FIR applied to the voltages) in the entire buffer, and window refers to the SNR using the maximum value in a small window around the the known injected impulse. "A pol" refers to simulations using the North-South-polarized dipoles and "B pol" refers to simulations using the East-West-polarized dipoles.

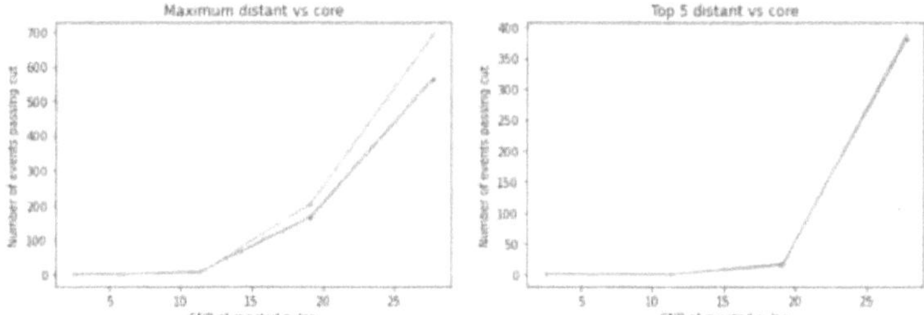

Figure 4.18: Left: Number of simulated events passing the criterion that the maximum SNR in the core be at least 2.5 times larger than the maximum SNR among distant antennas, plotted as a function of the SNR of the injected pulse. Right: Number of simulated events passing the criterion that the sum of the five largest SNRs in the core be at least 2.5 times larger than the sum of the five largest SNRs among distant antennas, plotted as a function of the SNR of the injected pulse. In both panels, blue points are for events with signals injected into the North-South dipole time series and orange points are for events with signals injected into the East-West time series. In all cases, the total number of events is 1000.

strategy and repeat the tests, and they are preliminary in the sense that their purpose is to reject the most serious RFI and promote a small subset of events for further inspection. These preliminary cuts are initially used to search for a power gradient from the core to the distant antennas are too strict. The next steps in searching for cosmic rays include exploring other versions of the preliminary cuts, using arrival directions to reject RFI from serious common sources, and applying a calibration in order to examine the power as a function of spatial location rather than using SNR as a proxy.

Path to Cosmic Ray Observing

The commissioning tests presented here show that a first cosmic ray detection with the upgraded OVRO-LWA will require the following:

1. **A search for event clustering in time and in arrival direction:** Monroe et al. [65] relied on event arrival directions to reject most RFI events in software by rejecting all events from a small fraction of the sky corresponding to known RFI source directions. Similarly, they used temporal clustering to reject RFI events from air planes (by searching for series of events tracing flight paths). Unlike this proof-of-concept analysis, my event filtering must eventually run in real time, and so I apply the computationally-simpler preliminary cuts described in this chapter, which reduce the number of events to inspect by a large fraction. The results in section 4.6 suggest that the preliminary cuts are effective at reducing the event rate by more than an order of magnitude (after relaxing the initially too-strict cuts so that cosmic rays are unlikely to be wrongly rejected), and the analysis is ready to proceed to direction reconstructions.

2. **Better cable delay estimates:** The current delay estimates from cable lengths are adequate for triggering on impulsive transients but need to be more precise in order to allow precise direction reconstructions. More precise delays can be solved for using a known source, which could be as simple as an impulsive RFI source on the horizon or a deliberately-radiated impulsive signal in the field (such as a barbecue sparker as used in [98]). For the distant antennas, these approaches are likely to work better than calibration on astronomical sources via the correlator infrastructure because the ionosphere can create delay differences on the order of 100 ns between core antennas and distant antennas. Direction reconstruction restricted to using only core antennas,

however, may be able to rely on delays from the correlator since the core array size is comparable to the size over which delays through the ionosphere are similar.

3. **Complete event filtering pipeline:** A complete event filtering pipeline requires code to compare the polarization and lateral power distribution to those expected from simulated air showers (see figure 4.1).

4. **Amplitude calibration:** The current software uses SNR as a proxy for event intensity. This proxy was a reasonable first approach given that the antennas have some differences in their gains and should have background noise mostly dominated by the same Galactic background for all antennas. An absolute calibration will allow the power in the impulses observed by each antenna to more directly be compared to each other, and will quantify the noise floor to better constrain the expected sensitivity to cosmic rays. Eventually, in addition to an absolute amplitude calibration, the total impulse response in the signal path should be measured so that it can be either deconvolved from the observed timeseries or used in a matched filter, for more precise pulse amplitude and fluence measurements.

5. **More data storage space:** Finally, larger datasets during commissioning will speed up the commissioning process. Monroe et al. [65] observed 8 cosmic rays in 40 hours. The upgraded system is designed to eventually run in real time, and so does not include a large amount of storage space on the dedicated cosmic ray computer. So far during commissioning, triggered snapshots from only a few hours of observing can be stored at one time, and must be deleted before attempting another observation. The ability to store enough data that one dataset should be nearly certain to contain a cosmic ray at the rate that Monroe et al. [65] observed would make the event filtering software development more straightforward. This larger storage will be possible in the very near future with a new storage computing cluster that is currently under construction at the OVRO-LWA.

Beyond first cosmic ray detections, the path to routine cosmic ray observing for a sample of thousands of cosmic rays for a composition analysis requires the following developments:

1. **Expanding the range of observing settings tested:** The detector settings (trigger and veto power thresholds and minimum number of antennas to trigger or veto) used in the tests in this chapter were chosen based on threshold scans aimed at a preliminary identification of settings that produced an acceptable detector dead time. The single-FPGA threshold scans could explore a wider range of minimum antenna numbers to trigger, which may further reduce the background trigger rate. Similarly, a fairly strict veto threshold is applied in the tests in this chapter, in the firmware settings and in the software processing. This strict veto could reduce sensitivity to events from the south, by vetoing some events that do not necessarily have a detection at the furthest northern veto antennas. The most recent observation had a whole-array trigger a factor of a few lower than the maximum acceptable rate, suggesting that there is leeway to run the veto with less stringent settings, potentially boosting the acceptance.

2. **Real time event filtering software:** The event filtering software must keep up with the rate that the events are observed. This goal is feasible because although the current version is too slow, the current software is a rapidly-written prototype in Python, and several ways to optimize it have already been identified. If these options are insufficient, porting it to a compiled language such as C will be a good option.

3. **Optimized resource-use for the bandpass filter:** The current firmware is very close to the limits of timing closure due to the resource use of the FIR filter, which also draws an amount of power that may stress the system for long-term observing. The likely path to reducing this FPGA resource use is to replace the bandpass FIR with a digital down-conversion followed by a simpler low-pass filter, for the same overall affect as the bandpass filter.

4. **Integrate monitor and control software:** The cosmic ray system software should be integrated into the main LWA control software for routine observing. This change requires changing only a few cosmic ray control software functions to send their commands to the SNAP2 boards via the same interface as the filterbank control software does, so that multiple observing programs can send queued commands to the FPGAs without conflicting with each other. The observations will not be queued since the system is designed for multiple simultaneous observing programs, but the various communications to the SNAP2 boards for monitor and control must go one at a time through

a single software interface to the SNAP2 boards. Cosmic ray composition analysis will require a long-term record of observing configurations and system up-times, and the recording of this system monitor data should also be incorporated into the overall LWA monitor and control software.

In summary, the developments outlined above provide a path to cosmic ray observing with the upgraded OVRO-LWA, based on the observing conditions demonstrated through the commissioning tests presented in this chapter.

Chapter 5

ENERGETIC PARTICLES IN THE MAGNETOSPHERES OF M DWARF STARS

Context and Motivation

This chapter explores the energetic particle environment of a magnetically-active M dwarf star. Dipolar magnetospheres can trap and accelerate particles in radiation belts, first observed for Earth and Jupiter. Pizzella [78] has tentatively detected cosmic rays from Jupiter up to nearly GV rigidities, and Kao et al. [56] have recently observed radiation belts around an ultracool dwarf star.

As discussed in Chapter 1, the magnetic field and size scale of an object determines the maximum-rigidity (momentum-to-charge ratio) particle the object could confine. Objects' locations in the parameter space of size and field strength (Hillas plot) show which objects are intriguing candidates for cosmic ray sources. Chapter 1 included a classic Hillas plot for the most-commonly-suggested cosmic ray source candidates (Figure 1.2). Figure 5.1 presents a Hillas plot with a focus on magnetized stars and planets (and a few other less-often-considered objects). Using its surface field strength as the characteristic field, the M dwarf flare star UV Ceti (the red circle on the plot) lies on a region that suggests it could contain protons well above a PeV and iron nuclei up to 100 PeV. All though such a simple criterion is likely an extreme upper limit, it does motivate further exploration of the high-energy particle environment of low mass stars.

This chapter has two parts. Part one introduces magnetospheric activity at the low-mass end of the main sequence and then presents radio observations of UV Ceti from 1–105 GHz. BL and UV Ceti are a nearby (2.7 pc) binary system with similar masses, spectral types, and rapid rotation rates, but very different magnetic activity (see section 5.1). UV Ceti's much stronger large-scale magnetic field may cause this difference, highlighting key unanswered questions about dynamo processes in fully convective objects. The first part of this chapter presents multi-epoch characterization of the radio spectrum of UV Ceti spanning 1 - 105 GHz, exhibiting flared emission similar to coronal activity, auroral-like emission analogous to planetary

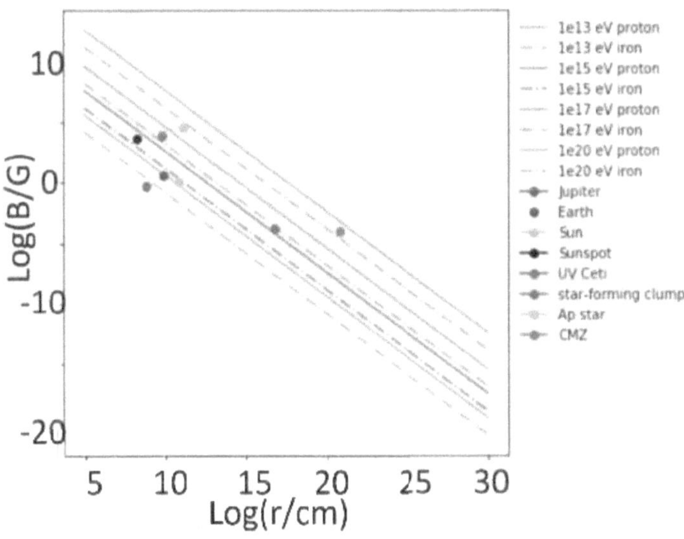

Figure 5.1: Hillas plot with an emphasis on stars and planets.

magnetospheres, and slowly-varying persistent emission. Radio observations are a powerful means to probe the role that the large-scale magnetic field of UV Ceti has in non-thermal particle acceleration, because radio-frequency phenomena result from both the activity of small-scale field features as well as large-scale auroral current systems. These observations reveal temporal variability at all bands observed, and a hint of rotational modulation in the degree of circular polarization up to 40 GHz. The persistent component of the emission is fairly constant from 1–100 GHz, making neither optically thick emission nor optically thin gyrosynchrotron from electrons with an isotropic pitch angle distribution likely. This part of the chapter includes discussion of the possibility of emission mechanisms analogous to Jupiter's radiation belts.

Part two includes calculations using two approaches to estimate rough upper bounds on the maximum-energy proton UV Ceti could accelerate. One approach follows a specific acceleration mechanism proposed for magnetized white dwarfs, adapting the calculation to UV Ceti. The other scales from planetary magnetospheres. From these calculations I conclude that 100 TeV protons in UV Ceti's magnetosphere may be plausible, although it is likely not quite a Pevatron. I finish the chapter by presenting some ideas for future directions to take this line of inquiry.

Part 1

RADIO EMISSION FROM THE MAGNETICALLY ACTIVE M DWARF UV
CETI FROM 1 GHz TO 105 GHz

[1]Kathryn Plant, [1]Gregg Hallinan, [2]Tim Bastian

[1]*Division of Physics, Mathematics, and Astronomy, California Institute of*
Technology, Pasadena, California 91125, USA

[2]*NRAO*

5.1   Introduction

M-dwarfs cooler than M3 or M4 are expected to be fully convective and thus must generate their magnetic fields with a different type of dynamo than the Sun [28], whose dynamo depends on the tachocline boundary layer between a convective outer region and a radiative core (e.g. [71]). Many M dwarfs exhibit far more intense magnetic activity than the present-day Sun, as evinced by: more frequent and more powerful flares (e.g. [40]), kiloGauss magnetic fields with large filling factors [54, 66, 33], higher ratio of X-ray and Hα luminosity to bolometric luminosity (e.g. [99]), and persistent nonthermal radio emission (e.g. [60]).

This magnetic activity persists for much longer than it does in the life-cycle of a sun-like star [97], consistent with slower spin-down timescales for fully-convective M dwarfs. Faster-rotating stars tend to be more magnetically active than slowly spinning stars. This relation results in a monotonic increase in the ratio of X-ray luminosity to bolometric luminosity ($\frac{L_x}{L_B}$) with an increase in the stellar Rossby number (ratio of rotation period to convective turnover timescale), a trend which saturates at the shortest rotation periods with $Log(L_x/L_B) \sim -3$ (for a summary see [99] and references therein).

One of the observable key differences between the dynamos in sun-like stars and fully convective M dwarfs is the prevalence of strong (up to a few kiloGauss) large-scale magnetic fields in M dwarfs, as revealed by Zeeman Doppler Imaging (ZDI; [33, 66]). It is becoming increasingly apparent that such large-scale magnetic fields can underlie differences in the magnetic activity observed. Large-scale magnetospheres can sustain stable current-systems that transport energy from the middle and outer magnetosphere into the lower atmosphere, resulting in auroral emissions (both radio and optical) from M dwarfs at the end of the main sequence as well as brown dwarfs (collectively termed ultracool dwarfs) [46, 47, 57]. The magnetic field configuration may also explain the bimodal distribution in spindown rates of M dwarfs [41].

The radio signature of auroral activity in ultracool dwarfs is the presence of periodic, highly circularly polarized radio emission, believed to be electron cyclotron maser emission (ECME) produced at either the local gyrofrequency, or the first harmonic (see e.g. [46, 57, 95, 101, 61]). It is detected from $\sim$ 20% of ultracool dwarfs ranging from spectral type M8 to T8. Although large optical flares are still observed from some ultracool dwarfs (e.g. [73], objects cooler than M7-M8 exhibit much less coronal X-ray and chromospheric Hα emission [24] than warmer spectral types, and this sharp difference suggests that auroral processes are the more significant magnetic activity for these cooler objects.

Most ultracool dwarfs observed to produce auroral emission also produce quiescent radio emission [77]. Recent VLBI observations of one such ultracool dwarf have revealed this radio emission to be produced by trapped energetic particles in a large-scale dipolar magnetic field ([56]), analogous to the radiation belts of Jupiter (e.g. [100]), further emphasizing the key role that large-scale magnetic fields play in defining the magnetic activity for these objects.

However, large-scale fields are not ubiquitous at the end of the main sequence, and very different field strengths and topologies are observed even in coeval fully-convective dwarfs of equal mass [59], with arguments put forward that the dynamo in rotating fully-convective objects may have two distinct stable states.

At 2.7 pc from Earth, the binary M dwarfs UV and BL Ceti are a well-characterized system for studying magnetospheres of fully-convective stars. With an orbital semimajor axis of 2.0584 AU, they do not directly interact. UV Ceti nearly identically resembles its less-active companion BL Ceti in age, spectral type (M6Ve and M5.5Ve, respectively [48]), mass ($0.1195 \pm 0.0043 M_\odot$ and $0.1225 \pm 0.0043 M_\odot$ [58]), and rapid rotation period (0.2269 ± 0.0005 day and 0.2432 ± 0.0006 day [20]), but UV Ceti exhibits more frequent flares and much brighter persistent X ray and radio emission.

Emission phenomena from UV Ceti include indicators of coronal and chromospheric activity typical of M dwarfs flare stars—frequent flares and saturated X-ray and Hα emission—but also include auroral activity more similar to ultracool dwarfs and planetary magnetospheres. As UV Ceti is the archetype M dwarf flare star, the initial detection of periodic, highly circularly polarized radio emission, consistent with auroral ECME emission ([61, 101, 95]), therefore came as somewhat of a surprise.

Different magnetic field configurations may interestingly underlie the difference in activity between UV Ceti and its twin BL Ceti. Zeeman Doppler Imaging confirmed that UV Ceti has a large-scale axisymmetric field and a rotationally-modulated field strength, while the magnetic field of BL Ceti is complex and non-axisymmetric [59].

In addition to exhibiting frequent and intense flares, M dwarf flare stars such as UV Ceti exhibit persistent coronal emission which bears similarities (intense X-ray emission and a non-thermal radio spectrum) to solar coronal emission during flares (see e.g. [34] and references therein). UV Ceti was one of the first main sequence stars detected at radio wavelengths, and the initial persistent detection of bright, non-thermal, coronal emission also came as a surprise since thermal Bremmsstrahlung emission dominates the solar corona [42]. This persistent emission, and the more recent discovery that ultracool dwarf auroral emitters tend also to exhibit persistent non-thermal radio emission (e.g. [77, 56]) highlight questions about the nature of the high-energy particle acceleration and particle-trapping required to power this persistent emission.

Gyrosynchrotron, synchrotron, and coherent emission processes make radio wavelengths a window for observing magnetic activity and high-energy particle acceleration in stars. Solar flares accelerate particles to mildly relativistic energies, up to MeV for electrons and 1 GeV/nucleon for atomic nuclei, and these particles produce X-ray and non-thermal radio emission in the sun's magnetic field during the flare (e.g. [34]). Magnetized planets and compact stellar remnants (magnetic white dwarfs and pulsars) accelerate high-energy nonthermal particles in stable large-scale current configurations.

Radio observations over a wide range of wavelengths can delineate the interplay of coherent and incoherent, thermal and non-thermal, and coronal and auroral emission processes. We present an analysis of archival observations of UV Ceti with the Karl G. Jansky Very Large Array (VLA) at 1–40 GHz and the Atacama Large Millimeter Array (ALMA) at 90–104 GHz. The VLA observations we include in this analysis investigate auroral and coronal processes over a range of stellar rotational phases. Section 2 presents details of the VLA and ALMA observations, section 3 presents the results, section 4 discusses possible emission scenarios, and section 5 concludes.

5.2 Methods

Observations

The VLA observed UV Ceti on January 8, 2011 in three bands between 1 and 25 GHz, and on January 28, 2011 in five bands between 2 and 40 GHz. The VLA was in B-North-C configuration for both observations. ALMA observed UV Ceti on July 23 2014 in band 3, from 90 to 105 GHz. Table 1 summarizes the observations. The total time elapsed between the start and finish of the set of observations, including time spent on calibrators, was 2 hours on January 8 2011 and 2.5 hours on January 28 2011. The VLA observed the different receiver bands sequentially, using all available antennas for each receiver. The option of making simultaneous observations in multiple bands using subarrays was not available yet at that time. The Ku band and S band images had reduced sensitivity because these VLA observations were made at the end of a major upgrade to the array, and the S band and Ku band receivers were not yet available on many of the antennas. ALMA observed a single receiver band and thus the entire range of observed frequencies was observed simultaneously.

Calibration

We used CASA (Common Astronomy Software Applications software package, [64]) to flag, calibrate, and image the VLA data, and to image the pipeline-calibrated data obtained from the ALMA archive. For the VLA observations, 3C48 served as both flux calibrator and bandpass calibrator. The phase calibrator is J0132-1654 at all frequencies, except for L band (2-4 GHz) which used J0116-2052. For the ALMA observations, Uranus served as the flux calibrator, J0132-1654 as the phase calibrator, and J0137-2430 as the bandpass calibrator.

Typical absolute flux density calibration uncertainty for the VLA is 5% at L–Ku bands and 10–15% at K and Ka bands [74]. For the L band phase calibrator J0116-2052, our flux measurement is within 6% of the VLA calibrator catalog value, consistent with the expected systematic uncertainty. The phase calibrator for the other VLA bands, J0132-1654 appears to be several times brighter than the fluxes in the VLA calibrator catalog. Long term monitoring of J0132-1654 at Ku band with the OVRO 40m telescope [84] shows that J0132-1654 underwent a large, slow brightening, and our flux measurement at Ku band agrees with the OVRO measurement from the following day (Sebastian Kiehlmann, private communication).

Imaging

We imaged the visibility data using the CASA tclean routine. We used Briggs 0 weighting, except with L band for which we used Briggs -0.5 to better mitigate sidelobes from a bright source. None of these archival observations include polarization calibrators, but since the VLA has left- and right-hand circularly polarized feeds, circular polarization (Stokes V) can be estimated without an additional calibrator. We imaged UV Ceti in Stokes V for all the VLA data, using imaging parameters that were otherwise similar to the Stokes I images. Only total intensity (Stokes I) images are possible for the ALMA data: ALMA's feeds are linearly polarized and only XX and YY correlation visibilities (no XY and YX correlations) were available in 2014 when the observation ocurred.

We measured the flux density of UV Ceti and the unresolved calibrators two ways: first by fitting a Gaussian source to the cleaned image via the CASA Gaussian fit tool and obtaining a best fit flux and statistical uncertainty from the fit, and secondly by measuring the peak pixel value on the source in an over-sampled image and the RMS in a region off the source. These two methods provided similar results. Flux results are reported in Figure 5.3.

At L band, a bright source 19.7 arcminutes from UV Ceti near the edge of the primary beam required self-calibration to reduce the contamination by its sidelobes, which otherwise would be inadequately removed by the CLEAN algorithm. We imaged nearly the full primary beam at L band, using three terms in the Taylor expansion. We used this first iteration image to create a model of the bright source, via the CLEAN algorithm, and then solved for a new phase-only calibration using that model.

Light Curves

We created light curves from the complex visibilities as follows. If the only source in the field of view is a point source at the phase center, then the real component of the visibility for any baseline is the flux of the source, and the imaginary component is zero. We used the CASA task fixvis to move the phase center to the position of UV Ceti (since the source is in a binary system, the original phase center was near but not centered on UV Ceti). At L and S band, where other sources are in the field of view of the primary beam, we used the CLEAN algorithm to create a sky model of all of the bright sources in the field of view except UV Ceti, and then subtracted these model visibilities from the measurement set. Subtraction of background sources was not necessary at higher frequencies due to the narrower primary beam and the spectrum of the sources. After subtracting the other sources, averaging the visibilities over all channels, spectral windows, and baselines creates a timeseries of the flux from UV Ceti. We obtained Stokes I and Stokes V time series from the separate time series made with visibilities from correlations of all of pairs of right hand circularly polarized feeds and separately all of the left hand circularly polarized feeds. We searched for variability on a range of timescales in each observation. We have not attempted to subtract BL Ceti from the visibilities, but we confirmed that the position of all flares is most consistent with the position of UV Ceti.

For additional confirmation of the slow variability observed with ALMA, we made single-scan images for each of the scans in the observation and confirmed from the point spread function in each single-scan image that the slow change was not due to a change in the quality of the phase calibration. Furthermore, we repeated the single-scan imaging for the phase calibrator and confirmed that the flux from phase calibrator did not vary during this time.

Astrometry

We measured the source position by fitting a Gaussian source to the cleaned images via the CASA Gaussian fit tool. The best-fit position of the source was consistent across all images.

As mentioned above, UV Ceti and BL Ceti are a wide binary with a 26-year period. We determined the expected positions of UV Ceti and BL Ceti in January 2011 (the time of the VLA observations) and July 2014 (the time of the ALMA observation) as follows. We used the coordinates and proper motions of UV and BL Ceti in the GAIA DR3 catalog to determine the position of the system barycenter at each epoch. We calculated the position angle and separation of the binary using masses and orbital parameters from [58] as input to the ephemeris calculator tool developed by Brian Workman[1]. The ALMA observation and the VLA observations at K band and Ka band have adequate resolution to resolve the binary, but only UV Ceti is detected in all the images except at K band on January 28 2011. At K band on January 28 2011, both components of the binary are detected. At lower frequencies, the synthesized beam is wider than the binary separation. The uncertainty on the centroid position, however, is smaller than the binary separation at all frequencies,

Band	Array	Frequencies	Polarization	Duration on Jan-8-2011	Duration on Jan-28-2011	Duration on Jul-23-2014
		[GHz]		[minutes]	[minutes]	[minutes]
L	VLA	1–2	Stokes I,V	29	–	–
S	VLA	2–4	Stokes I,V	–	20	–
C	VLA	4-5 & 7-8	Stokes I,V	29	14	–
Ku	VLA	13–14 & 16–17	Stokes I,V	–	25	–
K	VLA	19–20 & 24–25	Stokes I,V	24	18	–
Ka	VLA	30–31 & 39–40	Stokes I,V	–	18	–
Band 3	ALMA	90–94 & 102–106	Stokes I	–	–	37

Table 5.1: Summary of observations. Duration is the number of minutes spent observing UV Ceti.

allowing UV Ceti to be identified as the component responsible for the majority of the flux. Flares were localized to UV Ceti by imaging the time period of the flare separately, fitting a Gaussian to the source, and confirming that the position better coincided with the position of UV Ceti than BL Ceti.

5.3 Results

We detect UV Ceti at all epochs and observing bands. It exhibits persistent, slowly varying circularly polarized emission as well as flares (Figure 5.2). BL Ceti is only detected at K band on January 28 2011, and marginally detected at K band on January 8 2011. UV Ceti and BL Ceti were separated by 2.1 arcseconds in January 2011 and 2.3 arcseconds in July 2014 and thus, had BL Ceti been brighter, it would have been detected as a second source at least marginally resolved from UV Ceti in all the observations at frequencies above 10 GHz. Below 10 GHz, the separation of the components of the binary is smaller than the half-power width of the synthesized beam, but since the uncertainty on the position of the centroid of a point source is proportional to the width of the point spread function divided by the signal-to-noise ratio, we can fit the source position with sufficient precision to determine that UV Ceti, and not BL Ceti, dominates the observed flux at all observed frequencies.

Spectrum

Figure 5.2 shows the left-hand and right-hand circularly polarized light curves of all the VLA observations. Figure 5.3 shows the average flux of UV Ceti measured in each of the VLA and ALMA observations, plotted against the frequency of the band center of each observation. On January 8 2011, the flux from UV Ceti was fairly similar at each frequency band observed. On January 28 2011, UV Ceti was brightest during the C and S band observations. Since UV Ceti rotates with a period

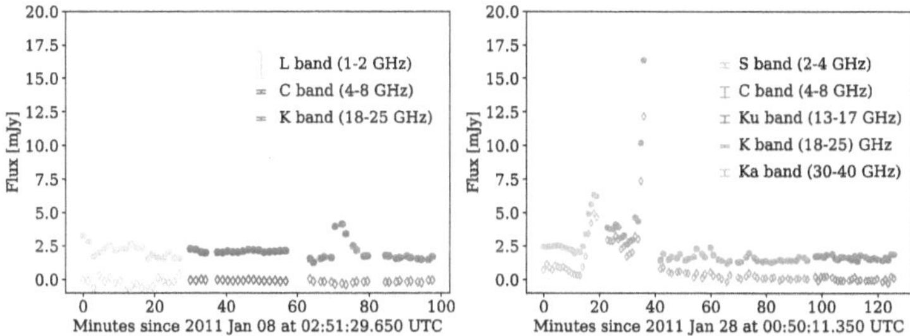

Figure 5.2: Light curve of UV Ceti for all of the VLA observations (left: January 8 2011; right: January 28 2011). Stokes I flux is shown with filled circles. Stokes V flux is shown with open diamonds. Color indicates the receiver band. The measured flux at each receiver band is shown, with no re-scaling to remove a spectral index. The time series are formed by moving the phase center to the position of UV Ceti and then averaging the visibilities over all baselines and frequency channels. The resulting light curves are binned to one data point per scan. Scans were one minute long for all except January 8 L Band and C Band, which were 90 seconds. Representative errorbars are shown in the figure legend. The typical uncertainty in the timeseries flux for each observation is estimated as $\sigma\sqrt{t_{\rm obs}/t_{\rm scan}}$, where $t_{\rm scan}$ is the scan length, obs is the observation length, and σ is the RMS of an off-source region of the entire-observation Stokes I image. Gaps in the timeseries were times during which the phase calibrator was observed, with the exception of a few gaps where scans were flagged due to contamination by RFI. For the L band and S band data, other bright sources in the field of view of the primary beam have been subtracted from the visibilities prior to calculating the light curve. This was not necessary at higher frequencies due to the narrower primary beam and the spectrum of the background sources.

of 5.45 hours [20], as well as flaring frequently, it is important to emphasize that the VLA observations in each band were conducted sequentially, not simultaneously, during each observation day. Figure 5.3 does not report instantaneous spectra, rather the flux measurements in each band are displayed on one plot for the purpose of providing a summary. In fact, the baseline component of the flux in Figure 5.2 suggests that a persistent component of the emission from UV Ceti may have a fairly flat spectrum. At Ka band, the upper and lower side bands are separated by nearly 10 GHz (the lower sideband is 29.487–30.511 GHz and the upper sideband is 38.487–39.511 GHz). Since the two sidebands are observed simultaneously, we can estimate a spectral index by imaging the two sidebands separately. The Ka band spectrum rises to higher frequencies, with a spectral index of 0.63 ± 0.27.

Millimeter emission is confidently detected on July 23 2014 with ALMA but is significantly fainter than all of the VLA observations (Figure 5.3). The millimeter emission has a spectral index -0.81 ± 0.36 (with the convention that the spectral index α is defined such that at frequency ν the flux F is proportional to ν^{α}), measured by separately imaging each of the four spectral windows in the ALMA observation and then fitting a power law to the measured flux vs frequency.

Figure 5.4 shows the circularly polarized fraction of flux from UV Ceti (Stokes V / Stokes I). On January 8 2011, UV Ceti is unpolarized or weakly left-hand circularly polarized depending on the band. On January 28 2011, UV Ceti is nearly completely circularly polarized at 6 GHz and has a significant component of right-hand circularly polarized emission at all frequencies observed except Ku band, which is nearly unpolarized. As with Figure 5.3, this plot of circular polarization fraction in each band is meant as a summary and does not imply spectral behavior, due to the non-simultaneous observations.

Temporal Variability: VLA

In the January 28 observation we detect intense, highly circularly polarized brightenings at C band and S band (see Figure 5.2). Both lightcurves are strongly circularly polarized both before and during the brightening events.

At K band, we detect two flares—one from each day. The flares rise sharply and decay approximately exponentially. The Stokes I timeseries of each flare are shown in Figure 5.5. Both flares have a spectrum rising to high frequencies, compared to flatter non-flare spectra. During the January 8 flare, the K band emission from UV Ceti has a spectral index 0.56 ± 0.17 compared to 0.008 ± 0.17 during the nonflare time. If the average flux from the non-flare part of the observation is subtracted from the total emission during the flare, then the spectral index becomes 0.99 ± 0.32. The emission remains 7–9% left-hand circularly polarized throughout the observation: during the non-flare time, the ratio of the Stokes V to Stokes I flux, V/I, is -0.09 ± 0.01. During the flare, V/I is -0.07 ± 0.02 without subtracting the persistant component, and is -0.052 ± 0.037 with the persistent component subtracted. These polarization fractions are consistent with no change during the flare.

On January 28 the spectral index of 0.31 ± 0.30 for the persistent emission steepens to 0.69 ± 0.32 during the flare, and the flare flux after subtracting the persistent component has a spectral index of 1.4 ± 1.28. The fraction of circular polarization

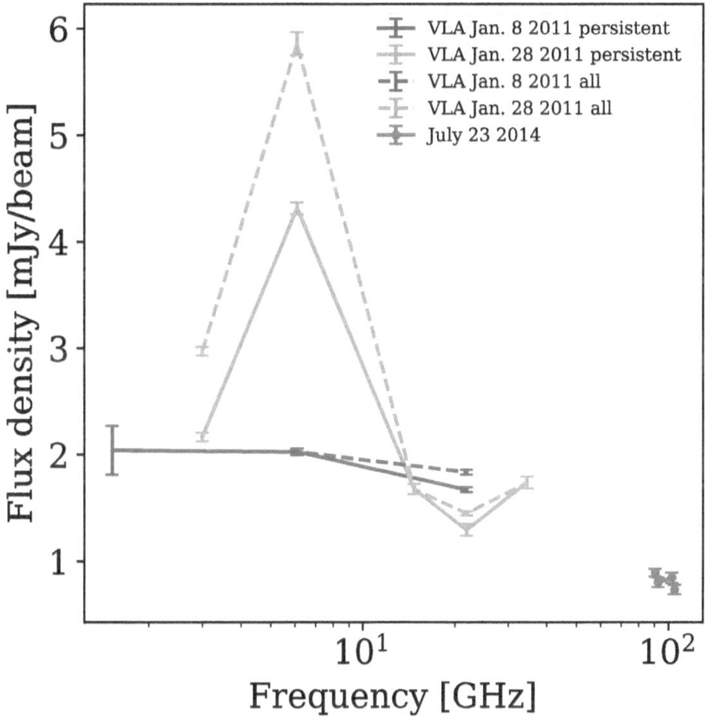

Figure 5.3: Flux density vs frequency measured with VLA and ALMA, with error bars showing statistical uncertainties. Typical systematic uncertainties for the VLA are on the order of 10–15%. Color indicates the different observing epochs. Lines connect the measurements at each band to make the groups of observations for each day easier to see. For the VLA observations, dashed lines, labelled 'all', show the flux at each band averaged over the full length of the observation. The solid lines, labelled 'persistent', show the flux averaged over the observation excluding times of flare or burst events. For the S and C band data, the large bursts have been excluded from the points labelled 'persistent', although the entire observation may be part of one extended burst. For the VLA data the flux shown is obtained from the peak pixel in the image of the source, while for the ALMA data it is the peak of a Gaussian fit to the source. The ALMA data is plotted with one point per spectral window.

increases from 0.15 ± 0.05 in the non-flare time to 0.25 ± 0.05 during the flare, and the flare emission with the non-flare average subtracted is even more strongly

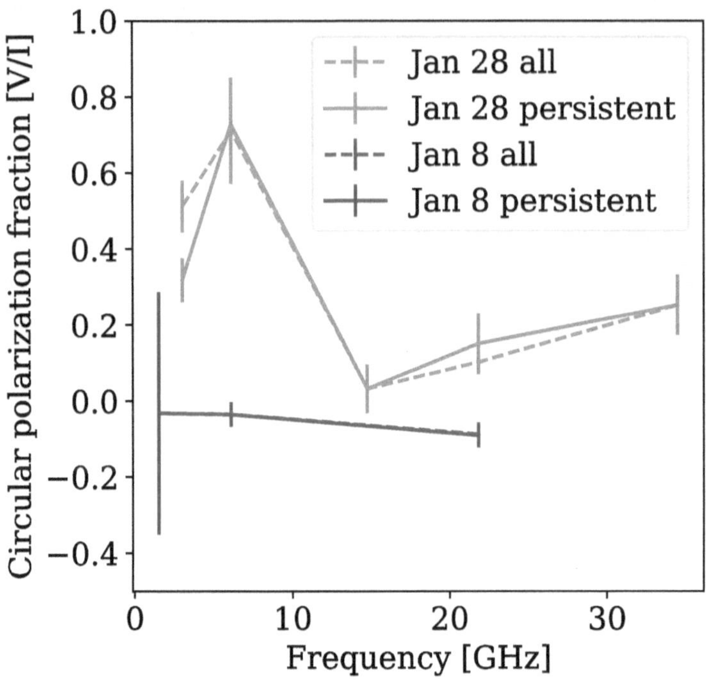

Figure 5.4: Circular polarization fraction measured in the VLA observations, measured as flux in Stokes V images divided by flux in Stokes I images, with errorbars showing statistical uncertainties computed by summing the uncertainties of the Stokes I and Stokes V images in quadrature. In images without a significant detection, we use the value in the Stokes V image at the position of UV Ceti determined from the corresponding Stokes I image. Color indicates the different observing epochs. Lines connect the measurements at each band to make the groups of observations for each day easier to see. For the VLA observations, dashed lines, labelled 'all', show the average over the full length of the observation. The solid lines, labelled 'persistent', show the average over the observation excluding times of flare or burst events.

circularly polarized with $V/I = 0.44 \pm 0.15$. Additionally, on January 28, the baseline non-flare flux shows a slow brightening during the observation.

The above estimates of the spectral index were obtained by imaging the upper and lower side bands separately for the time range of the flare. The uncertainty on each flux measurement is obtained from the off-source RMS in each image, and these uncertainties are propagated by Monte Carlo process to constrain the uncertainty on the spectral index. The quoted uncertainties V/I are estimated by taking the uncertainties from the off-source rms in each of the relevant images and analytically propagating the uncertainty through the polarization ratio calculation.

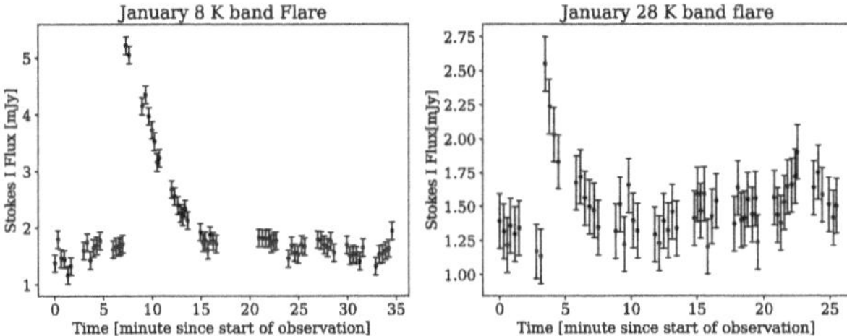

Figure 5.5: Left: Stokes I Flux density vs time for the K band January 28 observation. Right: Stokes I Flux density vs time for the K band January 28 observation. In both panels, the time series are smoothed to 20 s time resolution. Gaps are due to calibration scans and flagged data. The errorbars are $\sigma\sqrt{t_{\mathrm{obs}}/20s}$, where t_{obs} is the observation length, and σ is the RMS measured in an off-source region of the entire-observation Stokes I image.

Temporal Variability: ALMA

UV Ceti gradually brightened by 36% during the ALMA observation, as shown by Figure 5.6. The light curve in Figure 5.6 was made by imaging each of the five correlator scans separately, since there were fewer scans than in the VLA observations. The light curve made by averaging the visibilities with UV Ceti at the phase center shows the same brightening. The error bars in Figure 5.6 indicate statistical uncertainty due to thermal noise. These uncertainties are measured by computing the RMS of an off-source region of each image and were confirmed to be consistent with the expected sensitivity of this ALMA observation. Within the uncertainties, the measurement from the final scan is consistent with a continuation of the same gradual brightening and also consistent with the flux levelling off after the 36% brightening.

We confirmed that the observed slow brightening is not caused by a change in the phase calibration in two ways. First, we confirmed that the flux of the phase calibrator varies by only a few percent during this time. Second, if the phase calibrator is too far from UV Ceti, the quality of the phase calibration could change over time. We excluded this possibility by making a series of circular regions (in CASA region format) around UV Ceti, with increasing radius, and using the CASA imstat function to measure the total flux and peak flux density in each region. If the apparent brightening were due to phase errors causing an underestimate of the flux in earlier scans, then those scans would have more flux spread over a larger solid angle, which was not the case. Thus, UV Ceti's apparent brightening is likely intrinsic.

Using a higher time-resolution light curve computed by averaging the visibilities with UV Cet at the phase center, we searched for fast flares similar to the one-second millimeter flare from the M dwarf Proxima Centauri [63] and found none.

5.4 Discussion

In this section we discuss the possible emission mechanisms responsible for the combination of fast and slow phenomena we observed from UV Ceti. UV Ceti varies on such a range of timescales that "quiescent" may be a misnomer for any of the emission. Nontheless, time variability when clearly present, as well as brightness temperature, polarization and spectral index all inform our discussion of possible emission mechanisms. This section is organized in the following subsections: flares, auroral emission, and persistent emission.

Flares

The short flares observed at K band have the fast-rising exponential-decaying light curves of flares from magnetic reconnection events. The 5–45% circular polarization, and steepening flare spectra (spectral indices of 0.99 ± 0.32 during the January 8 and 0.69 ± 0.3 during the January 28 flare) are also consistent with gyrosynchrotron emission. Figure 5.9 shows the field strengths required for the range of harmonics of the cyclotron frequency that occur for gyrosynchrotron emission. The observed frequencies are consistent with emission regions of hundreds of Gauss, well above the surface in the corona of UV Ceti.

Auroral emission

Villadsen and Hallinan [95] and Zic et al. [101] observed periodic electron cyclotron maser emission from UV Ceti, which is the same process involved in auroral radio emission observed in Jupiter and ultracool dwarfs(e.g. [100, 46]). Villadsen and Hallinan [95] conclude that the coherent RCP bursts they observed from UV Ceti are electron cyclotron maser emission because the bursts are polarized in the sense

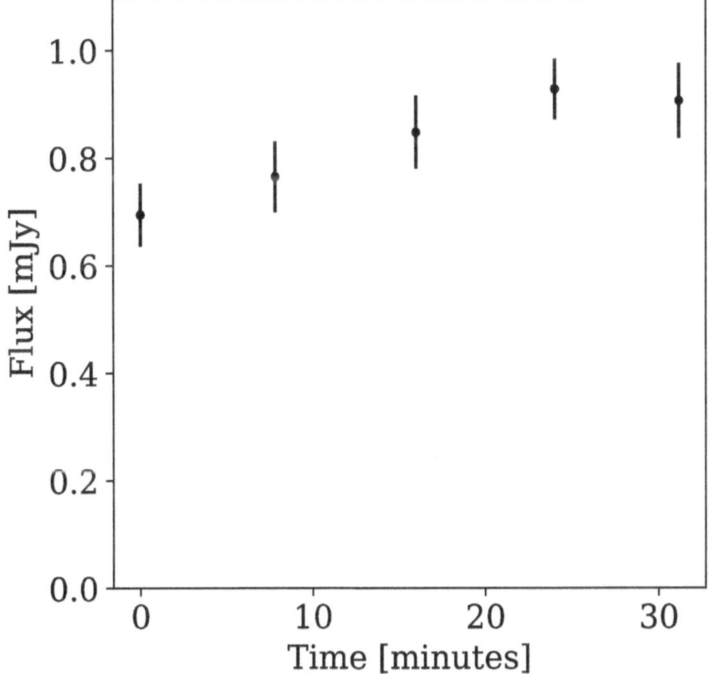

Figure 5.6: Flux density vs time during the ALMA observation. The time is displayed in minutes since the start of the observation. The flux values are the peak pixel value of an oversampled image, with one image made per correlator scan. Each of the first four scans is 6 minutes and 51.5 seconds long. The gaps between the scans are 30–45 second calibration observations. The final scan is five minutes and 16 seconds long. The error bars are the RMS in an off-source region of the image and are consistent with the expectation for thermal noise.

of the X mode, while plasma emission would be polarized in the sense of the O mode. The bursts we observe at C band and S band are polarized with the same sense as the bursts observed by Villadsen and Hallinan [95] and Bastian et al. [21] and thus consistent with being part of the same periodic electron cyclotron maser emission.

The observed frequencies of the ECM emission imply that the emission region spans field strengths of 0.5–1 kG if the emission is at the second harmonic of the cyclotron frequency, and 1-2 kG if it is at the first harmonic (see Figure 5.9). The magnetic field of UV Ceti has a strong dipole component (field strength 2 kG at the poles) as well as small-scale structure [59]. Considering only the dipole component, the field at radial distance r can be approximated as $B(r) = 2\,kG(r/r_*)^{-3}$ above the pole and a factor of two lower above the equator, where r is measured from the center of the star and r_* is the stellar radius measured by optical interferometry to be 0.159 ± 0.006 solar radii [58].

In the dipole field, field strengths of 0.5–1 kG occur at heights of 0.2–0.6 stellar radii above the surface of the star in the polar region and occur over the equator from 0.3 stellar radii down to the surface. If the emission is at the second harmonic, the required field strengths of 1-2 kG occur from 0.2 stellar radii down to the polar surface, and do not occur above the equatorial surface unless higher order components in the magnetic field are considered.

Persistent or slowly varying emission

This section discusses the persistent, slowly varying emission phenomena observed, including a potential observation of a variation in circular polarization with rotational phase (A), and a gradual brightening in the ALMA observation (B), and a discussion of the brightness temperature of persistent emission at each band.

(A) Slowly-varying circular polarization fraction

No flares were found at Ka band, but there is a gradual decline in the degree of polarization from 40% circularly polarized to 10% circularly polarized. This observation occurred following the intensely circularly polarized C band observation. January 8, 2011 and January 28, 2011 are close enough in time to estimate the relative stellar rotational phase between all of the VLA observations. Figure 5.7 plots light curves from all of the VLA observations at all frequencies vs relative rotational phase. As with Figure 5.2, the observed flux in each band is plotted without any scaling by the spectral index. The polarization fraction of the combined lightcurve intensely increases during the C band and S band January 28 observations, with all the other observations lying on the approximately exponentially-declining tail.

The Ka band observation shows a slow decline in the degree of polarization from 40% to 10% circular polarization over the duration of the observation, which occurred immediately after the lower-frequency ECM observation. Comparing the circular polarization fraction to stellar rotation phase (Figure 5.7) shows a peak at

the phase of the ECM followed by a slow decay. All the observations, regardless of
frequency and observing day, fall on this curve. Two interpretations are possible.
The higher-frequency observations on January 28 may have coincided with an unre-
lated flare. Interestingly, the observations from January 8 lie on approximately the
same curve, although the lightcurve has nearly flattened by that portion of rotation
phase. It is unfortunate that the only observations overlapping between days both
suffer from low signal to noise (due to fewer available antennas and data lost to RFI).
The other possibility is that the emission is rotationally modulated at all frequencies.
Simultaneous observations of UV Ceti at multiple frequencies over its full rotation
would clarify this picture.

If the Ka band emission were electron cyclotron maser emission, the implied magnetic field strengths exceed $10\,$kG for the fundamental frequency and $5\,$kG for the
first harmonic, for an observing frequency of $30\,$GHz. These field strengths exceed
the surface strength of the $2\,$kG dipole component.

In Jupiter's magnetosphere, electron cyclotron maser emission due to plasma from
Io results from stable large scale currents in a mostly dipolar field, but higher order
terms in the magnetic field are required in order to produce field strengths that
account for the highest frequencies at which the Jovian Io decametric radiation
is observed [80, 29]. UV Ceti does have significant small-scale magnetic fields
in addition to its strong dipole component, with a mean surface field strength of
$6.7\,$kG, suggesting that emission at low harmonics of the cyclotron frequency may
be possible even at the highest frequencies in the VLA observation.

Furthermore, at $30\,$GHz, the requirement that the emission frequency sufficiently
exceed the plasma frequency places less-stringent requirements on the density of the
emission region compared to ECM emission at S band and C band. $30\,$GHz emission
could easily propagate through typical M dwarf coronal densities of $10^{10.5}\mathrm{cm}^{-3}$ (see
list M dwarf coronal densities compiled in Villadsen and Hallinan [95]).

If the emission at Ka band is not related to the periodic ECM, gyroresonance and
gyrosynchrotron emission could be considered. Gyroresonance emission occurs at
the cyclotron frequency and its first few harmonics. As with the ECM scenario,
gyroresonance emission at Ka band would thus require field strengths corresponding
to the small-scale field structures near the surface. Gyroresonance emission can be
circularly polarized, but since gyroresonance emission is not beamed, the observed
change in polarization would require the majority of the polarized emission to be
confined in a region less than 10% of the stellar surface, in order to rotate out of

view over the course of the 18-minute observation.

Flares cause time-variable gyrosynchrotron emission, although these flares often accelerate electrons with a power law distribution, which produce typically only moderately circularly polarized emission. Gyrosynchrotron emission from a thermal population of electrons can be highly circularly polarized (>60%), with a strong, non-monotonic frequency dependence in the polarization fraction (Golay et al. 2023 submitted), although explaining the change in polarization over 18 minutes is more difficult for a thermal population.

Flares from the active corona may accelerate electrons that become trapped in radiation belts, making anisotropic pitch angle distributions important to consider. The presence of a strong dipole field [59] suggests that radiation belts similar to those recently observed in an ultracool dwarf [56] may be an important reservoir of radiating electrons. Fully modelling the spectrum of UV Ceti may require a 3D radiative transfer simulation such as that used by [67], who modelled emission from a population of electrons with a thermal component and a non-thermal tail, trapped in radiation belts around a star, and used their model to describe the active member of an observed RS CVN type binary system. They observed a change in polarization sense with frequency, but they did not consider pitch angle anisotropies. If the emission observed from UV Ceti at high frequencies is gyrosynchrotron, the emitting electrons may have an anisotropic pitch angle distribution, as has been used to explain radio emission from Jupiter's radiation belts [72].

(B) Millimeter emission

At millimeter frequencies, the flux density implies a brightness temperature of 6.5×10^5 K for an emitting region the angular size of the photosphere. The quiescent sun becomes optically thick to millimeter emission at the transition region at the top of the photosphere (see e.g. [12]), which is much cooler for UV Ceti, and thus the observed persistent millimeter emission is too bright to have an origin analogous to millimeter emission from the quiet sun. Gyroresonance emission is not likely at millimeter wavelengths because the implied field strengths are too large. If the emission is gyrosynchrotron emission, the field strengths implied are a few hundred to a few thousand Gauss depending on the harmonic number, which is consistent with the field expected for the stellar surface or low corona, and large emitting region-filling factors are required.

If the emission emanates from an extended large-scale magnetosphere, the brightness temperature could be 100 times lower (for a magnetosphere $\sim$ 10 stellar radii in diameter). Emission regions of size scales 1–10 stellar radii correspond to brightness temperatures of $6.5 \times 10^3 - -6.5 \times 10^5$ K which are within the 3–6 MK range of coronal temperatures estimated by [18]. However, depending on where exactly the coronal temperature lies in that range and how extended the millimeter emission is, the millimeter emission may be required to be optically thin.

The millimeter emission is slowly variable, brightening by 36% over 30 minutes. If this variability is caused by a brighter region of the star rotating into view on or near the stellar surface, then the minimum brightness temperature of that region, T_{region}, compared to the brightness temperature T_{baseline} prior to the brightening is

$$\frac{T_{\text{region}}}{T_{\text{baseline}}} = \frac{\Delta f}{f} \frac{P}{t}$$

where P is the rotational period (0.2269 days), t is the timescale of the brightening (30 minutes), and $\Delta f / f$ is the fractional change in flux (36%). Thus, using rotational modulation to explain the 36% brightening over 30 minutes would require a region with a brightness temperature at least four times the brightness temperature of the rest of the star to rotate into view. Alternatively, the variability could be intrinsic and related to a slow flare.

No fast flares similar to that observed for Proxima Centauri [63] were detected. However, if UV Ceti exhibited similar millimeter flares, these fast flares would occur too infrequently to expect a detection in an observation of the length we have in the ALMA archive. Millimeter flares on M dwarfs have been detected with timescales ranging from seconds [63, 62, 50] to tens of minutes or longer[2] [45, 68]. It is possible that the millimeter variability we observe on UV Ceti is part of a gradual flare. Future longer observations could identify flares or rotational variability, and polarimetry would help identify the emission mechanism.

(C) Brightness Temperature as a function of frequency

Figure 5.4 shows the brightness temperature of all of the bands observed, with flares and ECME bursts removed, plotted as a function of frequency. For reference, the dotted line plots the brightness temperature corresponding to a constant flux value (chosen as the mean flux of all the observations). The brightness temperature falls steadily with frequency, as the persistent component of emission is fairly similar

---

[2]Some observations ended before the end of the flare.

from 1–100 GHz. Although these are not simultaneous measurements, the similarity across all the bands makes it worth discussing the implications for the emission mechanism in a scenario where the persistent component in each observation does reflect a steady flux across all the days. If these points do approximate the spectral energy distribution of one unifying emission mechanism, the flux is too constant with frequency to be optically thick. In [35], a model of isotropic optically thick gyrosynchrotron emission in a dipolar magnetic field was used to describe the radio emission from the M dwarf component of a cataclysmic variable, where a radius-dependent electron population caused the spectrum to flatten and turnover above 10 GHz. This model, however, falls off too steeply at high and low frequencies and would not produce the flatness observed over the entire 1–100 GHz range. The flux is also too flat too be described by optically thin gyrosynchrotron emission from an isotropic distribution of electron pitch angle. A model analogous to Jupiter's magnetosphere may be required, with an anisotropic electron distribution confined to shells in the magnetosphere (e.g [72]).

5.5 Conclusions

In summary, UV Ceti exhibits slowly variable persistent emission at all the frequencies observed as well as flares that suggest a combination of auroral emission processes (ECM) and coronal emission processes (fast-rising exponential decay flares). In these observations, temporal variability is detected in every band observed. Fast-rising exponentially-decaying flares are observed at K band, consistent with gyrosynchrotron emission from flares due to small-scale magnetic field activity. Intensely circularly polarized flares are observed at C and S band, consistent with auroral ECME.

The circular polarization fraction appears to vary strongly with rotation phase even at Ka band. However, confirming this behavior would require a future observation at a wide range of frequencies simultaneously and covering a full rotation period. If the strong variability in the circular polarization fraction at Ka band is due to electron cyclotron maser emission, then the emission occurs in a region where the magnetic field is dominated by small scale field features that exceed 5 kG, rather than the dipole component of the magnetic field. Since electron cyclotron maser emission from Jupiter extends to frequencies that require involving higher-order magnetic field terms, this requirement may be reasonable for UV Ceti. Finally, the persistent (steady or slowly varying) emission maintains a fairly similar flux from 1–100 GHz, and makes energetic electrons trapped in a radiation belt with an

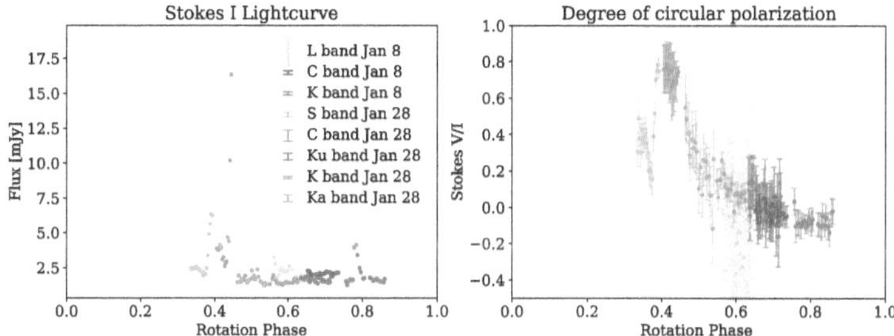

Figure 5.7: VLA lightcurves for all observations, plotted as a function of rotational phase. The zero phase is arbitrary but consistent for all observations. Lightcurves are binned to one data point per scan. Scans were one minute long for all except January 8 L Band and C Band, which were 90 seconds. **Left**: Stokes I light curves. Representative errorbars are shown in the figure legend, using the approximation that the noise in the Stokes I and Stokes V images from the same visibilities are the same. The typical uncertainty in the timeseries flux for each observation is estimated as $\sigma \sqrt{t_{\mathrm{obs}}/t_{\mathrm{scan}}}$, where t_{scan} is the scan length, t_{obs} is the observation length, and σ is the RMS measured in an off-source region of the entire-observation Stokes I image. **Right**: Degree of circular polarization. The colors in the legend from the right panel apply to the left panel, but the errorbars are displayed for each point because the uncertainty on the polarization fraction, $\sigma_{\mathrm{V/I}}$, is obtained by propagating the uncertainty on the flux, such that $\sigma_{\mathrm{V/I}} = \frac{\sigma}{f_I}\sqrt{1 + \frac{f_V}{f_I}}$, where f_V and f_I are the Stokes V and Stokes I fluxes, respectively.

anisotropic pitch-angle distribution an interesting possibility.

KP thanks Ivey Davis and Jackie Villadsen for many useful discussions about flare stars, and Dillon Dong for immensely helpful conversations about VLA data reduction and CASA. KP was supported by an NSF GRFP fellowship for part of the time that this research was conducted. To check the flux of the phase calibrator, this research has made use of data from the OVRO 40-m monitoring program (Richards, J. L. et al. 2011, ApJS, 194, 29), supported by private funding from the California Insitute of Technology and the Max Planck Institute for Radio Astronomy, and by NASA grants NNX08AW31G, NNX11A043G, and NNX14AQ89G and NSF grants AST-0808050 and AST- 1109911.

Part 2

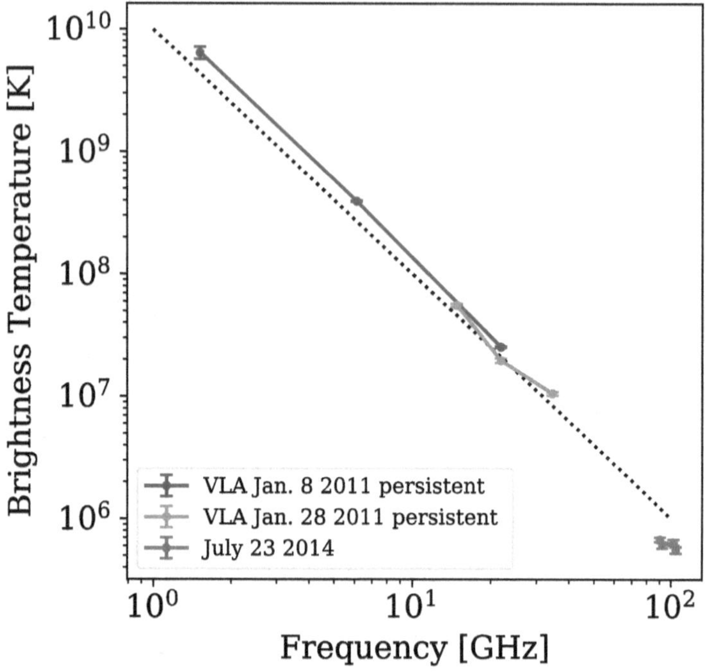

Figure 5.8: Brightness temperature for the persistent component of emission at all bands for which persistent emission was observed (the January 28 S band and C band data is excluded as the ECME burst dominated the observation). The brightness temperature is calculated for the angular size of the photospheric disk, and would be lower if the emitting region encompasses an extended magnetosphere. Color indicates the different observing epochs. Lines connect the measurements at each band to make the groups of observations for each day easier to see. For the ALMA data, each one point is plotted for each spectral window. Errorbars are calculated from the statistical uncertainties in the images, and are typically smaller than the marker size. The brightness temperature is a factor of 2–3 lower if the angular size of the VLBI observation from [23] is used, and potentially 100 times lower for an extended magnetosphere. The dotted line plots the brightness temperature for constant flux (using the mean value of the flux), emphasizing that the flux is similar at all frequencies observed. For the VLA data the flux shown is obtained from the peak pixel in the image of the source, while for the ALMA data the flux is the peak of a Gaussian fit to the source.

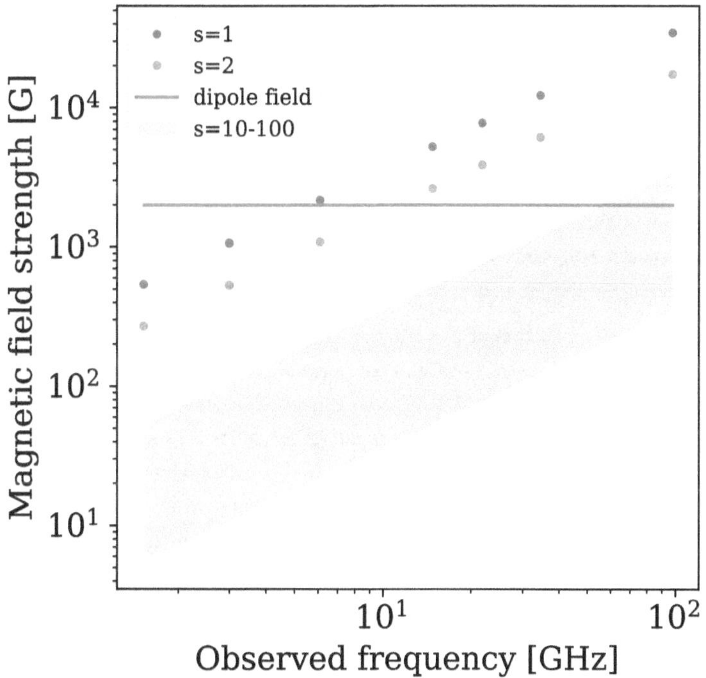

Figure 5.9: Magnetic field strengths required for emission at the observed frequency to be the specified harmonic s of the cyclotron frequency. Fields for the fundamental and second harmonics ($s = 1$ and $s = 2$) are plotted for the frequencies of the band centers of each of VLA and ALMA observations. The blue shaded region encompasses the range of harmonics applicable to gyrosynchrotron emission. The green line marks the polar surface field strength of the dipole component of UV Ceti's magnetic field.

COULD THERE BE GALACTIC PEVATRONS AT THE END OF THE MAIN
SEQUENCE?
PREDICTIONS FOR THE UPPER LIMITS OF HADRON ACCELERATION IN
M DWARF MAGNETOSPHERES

In the second part of this chapter I explore the prospects of high-energy cosmic ray acceleration in M dwarf magnetospheres by examining two particle acceleration energy limits that have been proposed for other objects with strong dipolar magnetic fields (white dwarfs, Jupiter, Earth's Van Allen belts) and scaling these limits to UV Ceti's magnetosphere.

Magnetized regions can trap charged particles whose energy is not so high that the Larmor radius of the particle exceeds the size of the region. This observation is known as the Hillas Criterion (see Chapter 1) and provides a simple assessment of the plausibility of various objects as candidate cosmic ray accelerators. This criterion is met for an ultra-relativistic particle of energy E and charge q when the rigidity $\mathcal{R} \equiv E/q \leq rB$. Here, B is the magnitude of the magnetic field and r is the length scale of the region over which this field exists.

The simplest application of this criterion suggests that the magnetosphere of the rapidly-rotating M dwarf UV Ceti could potentially contain cosmic ray protons with energies up to roughly 20 PeV, placing m dwarfs among the candidates for the elusive Galactic Pevatrons. Here I use the stellar radius 9.74×10^9cm for a size scale and the mean surface field strength of 6.7 kG measured by [59]. The Hillas criterion, however, does not imply that a suitable acceleration process exists to reach the upper limit particle energy. Furthermore, the simple limit calculated above does not consider how the field strength falls off with distance from the surface. In the sections that follow, I explore more concrete upper limits by analogy to the magnetospheres of compact stellar remnants (Section 5.6) and planets (Section 5.7). Section 5.8 concludes the chapter.

5.6   Analogy to Pulsars and White Dwarf Magnetospheres

Magnetized, compact stellar remnants such as pulsars and magnetized white dwarfs (polars) accelerate high-energy particles in strong magnetic fields. Arons and Scharlemann [16] (and associated papers in that series), calculated the change in electric potential above the polar cap of a neutron star. In *"High-energy emission of fast rotating white dwarfs"*, Ikhsanov and Biermann [52] adapt that calculation (removing general relativity terms) to estimate the maximum energy of particles accelerated in the magnetosphere of a rapidly rotating white dwarf with a strong dipolar magnetic field. Here, I follow the calculation from [52] and adapt it to UV Ceti.

Calculation of Maximum Particle-Energy

Ikhsanov and Biermann [52] start from the following equation for the component of the electric field parallel to the magnetic field in the polar cap regions, as a function of radial distance s from the surface of the star:

$$
E_\| = E_{AS} \times
\begin{cases}
\dfrac{s}{\Delta R_p}, & 0 \le s \le \Delta R_p \\[2mm]
1, & \Delta R_p \le s \le R \\[2mm]
\sqrt{2R/r}, & s > R
\end{cases}
$$

where

$$
E_{AS} = \frac{1}{8\sqrt{3}}\left(\frac{\Omega R}{c}\right)^{5/2} B_0
$$

The radius of the star is R, ΔR_p is the size of the polar cap, $\Omega = \frac{2pi}{P}$ is the angular rotational rate for period P, c is the speed of light, B_0 is the magnetic field at the poles. Scharlemann et al. [86] define s as the distance from the stellar surface along a magnetic field line. [3]

They find the electric potential $\phi(r)$ as a function of radial distance by integrating the electric field along a magnetic field line, from the stellar surface $r = R$ (approximately a radial path here for the polar region).

$$
\phi(r) = \int_R^r E_\| ds = \int_R^r E_{AS}\sqrt{2R/s}\, ds = E_{AS}\sqrt{2R}\int_R^r s^{-1/2}\, ds
$$

$$
= 2\sqrt{2R}E_{AS}(r^{1/2} - R^{1/2}) = 2\sqrt{2}E_{AS}R((r/R)^{1/2} - 1)
$$

The maximum energy of particles accelerated in this potential is the product of the charge of the particle and the potential evaluated at the end of the acceleration region, a radial distance l_0 from the center of the star.

Ikhsanov and Biermann [52] express the result scaled for the radius, rotation rate, and magnetic field of their white dwarf. My final result disagrees with theirs by a factor of 25, and so I show the details of this calculation in the appendix.

Applying this approach to UV Ceti

Below I present the maximum particle energy using the radius, magnetic field, and rotation rate of UV Ceti. I present the result using the scale factor I obtained (4.9×10^{12} eV) as well as the scale factor from [52] for comparison (2×10^{11} eV). The length of the accelerating region is uncertain, and Ikhsanov and Biermann use

---

[3] Although [52] refer to r as the distance above the surface, I believe that they meant s since they define $r = R + s$.

the polar cap size as a lower bound ($l_0 = R + \Delta R_p$) and the radius of the light cylinder $l_0 = R_{lc} = c/\Omega$ as the upper bound. Although the light cylinder is not as likely to be a useful upper bound for UV Ceti, I include it for comparison. I also evaluate the maximum particle energy for an intermediate acceleration length of three stellar radii, similar to the size scale of the synchrotron lobes. I use $2\,\mathrm{kG}$ as the strength of the dipole component of the magnetic field, and it should be noted that the mean surface field strength is several times larger [59]. It would be interesting to compare the results using a value that takes out the contribution from higher-order terms.

For UV Ceti, I use:

$R = 0.14\,R_{\mathrm{sun}} = 9.74 \times 10^9\,\mathrm{cm}.$

$P = 5.4456\,\mathrm{hr}$

$B_0 \sim 2\,\mathrm{kG}$

The dipole moment is $\mu = \frac{1}{2}B_0 R^3 = 9.3 \times 10^{32}\,\mathrm{G\,cm^3}.$

The radius of the light cylinder is $R_{LC} = \frac{c}{\Omega} = \frac{cP}{2\pi} = 9.354 \times 10^{13}\,\mathrm{cm}.$

The polar cap size is $\Delta R_p = R\sqrt{\frac{\Omega R}{c}} = R\sqrt{\frac{R}{R_{LC}}} = 9.94 \times 10^7\,\mathrm{cm}.$

The table below summarizes the results using these values.

pre-factor	acceleration length [cm]	maximum energy [eV]
2×10^{11}	$R + \Delta R_p = 9.8 \times 10^9$	2.6×10^{11}
5×10^{12}	$R + \Delta R_p = 9.8 \times 10^9$	6.5×10^{12}
2×10^{11}	$R_{lc} = 9.354 \times 10^{13}$	5.0×10^{15}
5×10^{12}	$R_{lc} = 9.354 \times 10^{13}$	1.24×10^{17}
2×10^{11}	$3R = 2.922 \times 10^{10}$	3.8×10^{13}
5×10^{12}	$3R = 2.922 \times 10^{10}$	9.4×10^{14}

Thus, this emission mechanism could produce $\sim$TeV electrons and protons in the most conservative scenario (minimum acceleration length). If the acceleration region extends to several stellar radii, acceleration to $\sim 1\,\mathrm{PeV}$ may be possible.

Does UV Ceti satisfy the assumptions for using this model for the electric field?
Ikhsanov and Biermann's calculation of the electric potential required the assumption that material in the magnetosphere does not affect the electric potential at the surface. The equation for the electric field requires that the star can be approximated as a surface surrounded by vacuum. They state that this assumption can be checked

by comparing the size of the scale height of the atmosphere to the size of the polar cap.

For a hydrogen atmosphere, the pressure scale height is

$$h = \frac{kTR^2}{m_p G M}$$

where k is the Boltzmann constant, T is the temperature, m_p is the mass of a proton, G is the gravitational constant, and M is the mass of the star. For UV Ceti, I use the photosphere temperature 2900 K (following [59]) and mass $0.102\,M_{\text{sun}}$.

These values give $h = 1.687 \times 10^6 \text{cm}$, which is much less than the size scale of the polar cap, allowing the above approximation. For the mass, radius and temperature of Ikhsanov and Biermann's white dwarf, I get a scale height of $1.5 \times 10^4 \text{cm}$ but they get $2.5 \times 10^4 \text{cm}$. A more detailed model could test the assumption used here that material in the magnetosphere does not modify this electric field.

Applying this approach to Jupiter

I repeated these calculations for Jupiter, as summarized below.

$M = 1.898 \times 10^{30} g$

$R = 7.15 \times 10^9 \text{cm}$

$B_0 = 8.34 G$

$\mu = 1.55 \times 10^{30} \text{Gcm}^3$

$P = 9.925 hr$

$T = 170 K$

$R_{LC} = \dfrac{cP}{2\pi} = 1.68 \times 10^{15} cm$

$\Delta R_p = R\sqrt{R/R_{LC}} = 1.475 \times 10^7 cm$

$h = 5.42 \times 10^6 cm < \Delta R_p$

For the result below, I used my calculation of the prefactor and set the minimum l_0 as $R + \Delta R_p$ and the maximum as the radius of the inner radiation belts, which is $5R$.

For the inner l_0, the maximum energy would be 0.043 eV, and for $l_0 = 5R$ the maximum energy would be 52 eV. Thus, Jupiter's high-energy particles do not originate from this mechanism. In the next section I scale a model based on planetary radiation belts to UV Ceti.

5.7 Scaling from Solar System Radiation Belts I: Scaling from Pizzella 2018

In *"Emission of cosmic rays from Jupiter: magnetospheres as possible sources of cosmic rays"* Pizzella [78] estimates upper limits to Jovian cosmic ray energies by scaling from the Earth's Van Allen belts and furthermore shows that the cosmic ray particle flux measured at Earth increases when the relative positions of Earth and Jupiter in their orbits places Earth on the path that cosmic rays from Jupiter would follow in the magnetic field of the solar wind. To estimate the maximum energy of cosmic rays from Jupiter, they start with a particle-trapping criterion (similar to the Hillas criterion) that gives an upper limit energy based on the size and field strength of the magnetosphere. They compare this upper limit, evaluated for the Earth's magnetosphere, to the actual highest energy of particles observed in the Van Allen radiation belts. Then, they calculate what the highest energy of cosmic rays from Jupiter would be if Jupiter's magnetosphere possesses an acceleration mechanism that reaches the same fraction of the trapping-criterion upper limit as was found for the Van Allen belts. The advantage (and in some ways disadvantage) of this approach is that it avoids specifying an acceleration mechanism.

Pizzella uses a cosmic ray confinement criterion (see Appendix at the end of this chapter) that results in a maximum cutoff energy

$$E_{\max} = \frac{\epsilon q \mu}{3 r^2}$$

where ϵ is a dimensionless parameter, r is the size of the accelerator region, μ is the magnetic dipole moment, and q is the charge. I am working in CGS units and so this equation lacks the factor of c from Pizzella's equation 2, which is in SI. Substituting $B = \mu r^3$ shows that this criterion differs from the Hillas criterion $E_{\max} = rBq$ (that I used in the introduction) only by the factor of $\epsilon/3$.

Pizzella then estimates the maximum cosmic ray energy from Jupiter by scaling from Earth's Van Allen belts as follows. The observed maximum particle energy in the Van Allen belts (Pizzella uses 500 MeV) is somewhat less than the cutoff energy obtained with the above criterion. They then calculate what the maximum cosmic ray energy from Jupiter would be if Jupiter's magnetosphere accelerates cosmic rays to the same fraction of that cutoff limit as the Van Allen belts do. Thus, they calculate

$$E_{\max} = E_{\text{Earth\_Max}} \frac{\mu}{\mu_{\text{Earth}}} \left(\frac{r}{r_{\text{Earth}}} \right)^{-2}$$

where $E_{\text{Earth\_Max}}$ is the maximum particle energy in the Van Allen belts, μ_{Earth} is the magnetic dipole moment of Earth and r_{Earth} is the radius of the Earth. Using $r_{\text{Earth}} = 6.37 \times 10^8$ cm, $\mu_{\text{Earth}} = 8 \times 10^{25}$ G cm^{-3}, $E_{\text{Earth–Max}} = 500$ GeV for Earth's magnetosphere, and for UV Ceti $r = 9.7 \times 10^9$ cm and $\mu = 9.3 \times 10^{32}$ G cm^{-3} gives $E_{\max} \sim 25$ TeV.

5.8 Conclusions

In conclusion, these two different energy-limit estimates both suggest that protons reaching $\sim 10^{13.5}$ eV in UV Ceti's magnetosphere are plausible. Although UV Ceti is likely not a Pevatron based on these limits, the polar-cap acceleration model may allow PeV protons for certain values of the parameters. If UV Ceti possess radiation belts, multiple acceleration mechanisms may be in effect. If particles accelerate over the poles similar to the White Dwarf model discussed above, these particles could continue accelerating via a mechanism (such as co-rotation breakdown in a current sheet) that applies to Jupiter's radiation belts. Furthermore, an interesting difference between UV Ceti and the magnetospheres of white dwarfs and solar system planets is that UV Ceti possesses an active corona, whose flares inject mildly relativistic particles into the magnetosphere, which may be a starting point for other acceleration processes.

Future steps to further assess the importance of M dwarfs as candidate Galactic cosmic ray sources include calculating the total contribution these magnetospheric accelerator models could account for in the locally-observed flux, and secondly predicting the associated gamma rays flux. High energy protons colliding with material around a star would produce gamma rays via pion decay. Song and Paglione [90] did not detect gamma ray emission from M dwarfs in a search stacking Fermi maps at the positions of a sample of M dwarfs. They did however claim a tentative detection of a single ultracool dwarf as a pulsed gamma ray emitter. Considering the variety of field configurations in M dwarfs (e.g. [59]), the M dwarf gamma ray search should perhaps be repeated for a sample restricted to those known to have strong large-scale, dipolar fields.

To follow up the potential observation of rotationally-modulated circular polarization discussed in the first part of this chapter, I led a successful proposal for 18 hours of VLA time in order to observe UV Ceti over a wide frequency range simultaneously. Sub-array mode allows the VLA to be configured so that different receivers are used simultaneously by different subsets of antennas, such that a wider range of frequencies can be observed simulataneously for bright sources such as UV Ceti that do not require the sensitivity of the full VLA for a detection. Of the allocated 18 hours, 15.5 hours of data in subarrays at C, X, and Ku band were obtained (less than the full allocation due to large over-subscription during the required LST range and due to sub-array mode requiring technical work until later in the semester). These observations provides simulataneous frequency coverage from 4–18 GHz contiguously over one full rotation period and several shorter observations within a timerange that can be rotation-phase-connected to the full-period observation. This dataset will allow a search for a high-frequency cutoff to the ECME and further elucidate the emission mechanisms responsible for the persistant component. Even higher frequencies were not available due to limited time with weather suitable for K–Ka band observing, but would be interesting to observe in the future.

K.P. thanks Noemie Globus for pointing me on a path to the energy limits calculations after I had the idea to put UV Ceti on a Hillas plot.

5.9 Appendix: White Dwarf Maximum Energy Calculation

Below, I show my attempt to derive from equation 9 in [52] (the electric field) their equation 10 (maximum particle energy evaluated for the white dwarf). I show this calculation in detail because my result differs from theirs by a factor of ~ 25.

For a charge q, the maximum particle energy E is:

$$E = q\phi(l_0)$$

$$= q \times 2\sqrt{2}E_{AS}R((l_0/R)^{1/2} - 1)$$

$$= q \times 2\sqrt{2}\frac{1}{8\sqrt{3}}(\frac{\Omega R}{c})^{5/2}B_0 R((l_0/R)^{1/2} - 1)$$

$$= q \times \frac{\sqrt{2}}{4\sqrt{3}}(\frac{\Omega R}{c})^{5/2}B_0 R((l_0/R)^{1/2} - 1)$$

$B_0 = \frac{2\mu}{R^3}$ where μ is the dipole moment of the star.

$\Omega = \frac{2\pi}{P}$ where P is the period. Making these substitutions,

$$E = q \times \frac{\sqrt{2}}{4\sqrt{3}}(\frac{2\pi R}{Pc})^{5/2}\frac{2\mu}{R^3}R((l_0/R)^{1/2} - 1)$$

$$= q \times \frac{4}{\sqrt{3}}(\frac{\pi}{c})^{5/2}P^{-5/2}\mu R^{1/2}((l_0/R)^{1/2} - 1)$$

$$= q \times \frac{4}{\sqrt{3}} (\frac{\pi}{c})^{5/2} (33s)^{-5/2} (10^{34.2} \text{Gcm}^3)(10^{8.8}\text{cm})^{1/2}$$

$$\times (\frac{P}{33s})^{-5/2} \frac{\mu}{10^{34.2}\text{Gcm}^3} (\frac{R}{10^{8.8}\text{cm}})^{1/2}((l_0/R)^{1/2} - 1)$$

$$= q \times 4.9443 \times 10^{12} \text{V}(\frac{P}{33s})^{-5/2} \frac{\mu}{10^{34.2}\text{Gcm}^3} (\frac{R}{10^{8.8}\text{cm}})^{1/2}((l_0/R)^{1/2} - 1)$$

For a proton or electron, this maximum energy is:

$$E = 4.9443 \times 10^{12} \text{eV}(\frac{P}{33s})^{-5/2} \frac{\mu}{10^{34.2}\text{Gcm}^3} (\frac{R}{10^{8.8}\text{cm}})^{1/2}((l_0/R)^{1/2} - 1)$$

This matches Ikhsanov and Biermann equation 10 except that they obtain 2×10^{11}eV where I have 4.9443×10^{12}eV.

5.10 Appendix: Reproducing Pizzella's Calculation for Jupiter

Here I show how Pizzella's equation 2 (criterion for E_{max}) can be derived from their equation 1:

$$(\frac{R_L}{B})(\frac{dB}{dr}) < \epsilon$$

where R_L is the Larmor radius, B is the magnitude of the magnetic field strength, and r is the radial distance from the center of the object.

The dipole field of an object with dipole moment μ is $B(r) = \mu r^{-3}$ and thus $\frac{dB}{dr} = -3\mu r^{-4}$. In the relativistic limit, $R_l = \frac{E}{qB}$ Making these two substitutions into the right hand side of equation 1 yields:

$$(\frac{R_L}{B})(\frac{dB}{dr}) = \frac{-3\mu r^{-4}\frac{E}{qB}}{\frac{\mu}{r^3}} = \frac{-3E}{qBe} = \frac{-3Er^2}{q\mu}$$

Setting the absolute magnitude of this to be less than ϵ and solving for E gives

$$E < \frac{\epsilon q\mu}{3r^2}$$

I took the absolute magnitude in order to arrive at their result. Using the values of parameters for Jupiter that I used in section 1, I get a maximum energy of 75 GeV which is close to their result of 70 GeV. The value they give for Jupiter's dipole moment has a typo, making it an order of magnitude too small, but is not the value they used to get 70 GeV.

Chapter 6

CONCLUSIONS AND FUTURE DIRECTIONS

6.1 Summary of this book

This book has described the design and implementation of a radio-only cosmic ray detector built to for the purpose of sensing air showers from the highest-energy Milky Way cosmic rays. At the OVRO-LWA, access to search the raw ADC timeseries on field programmable gate arrays makes a radio-only self-triggered cosmic ray detector possible in parallel with the other array activities. Chapter two described the OVRO-LWA and its recent major upgrade. This chapter highlighted the aspects that are most relevant to cosmic ray detection, including the array layout, bringing analog RF signals from each antenna to a central location for digital signal processing, precise signal timing synchronization, and a system for reading out snapshots of data from the entire array at once.

Chapter 3 described the FPGA firmware for the cosmic ray search, which consists of three main parts: (1) a system to search for coincident pulses among subsets of antennas in the core array, with distant antennas serving as an RFI-veto; (2) a data buffer that transmits snapshots of data over Ethernet; and (3) a state-control system that manages whether the FPGAs read out data over Ethernet or write new data to the buffers. This system accomplishes a roughly 4% readout dead time for a 50 Hz trigger rate.

Chapter 4 discussed the process of commissioning the cosmic ray detector, including characterizing the RFI background and the event rates during operation of the cosmic ray trigger. This chapter also presented the first steps toward classifying the impulsive transients that are recorded, and outlined the next directions for searching for cosmic rays.

Chapter 5 examined the energetic particle environment of an M dwarf magnetosphere. This chapter presented an analysis of archival VLA and ALMA data from 1–105 GHz for the nearby flare star UV Ceti as well as calculations on the prospects of cosmic ray acceleration in this magnetosphere. I found that a combination of coronal and auroral magnetic activity took place during these observations, consistent with the unusual combination of phenomena previously-observed in this star. Activity that had not been previously documented consisted of a dramatic change

in the circular polarization fraction at high (30 GHz) frequencies, closely following
the known low frequency electron cyclotron maser emission (ECME) bursts. For
future work, I successfully proposed for VLA time to follow up this observation
with a VLA observation in sub-array mode, with which the VLA observed UV Ceti
in three bands simultaneously over an entire rotation period, to confirm whether a
high-frequency phenomenon accompanies the period ECME.

6.2 Future developments to look forward to

Near-future goals for the OVRO-LWA

I will continue commissioning the OVRO-LWA cosmic ray detector as an NPP
fellow at JPL. The immediate next work toward this goal is as follows:

- Finalize an event classification pipeline that identifies cosmic rays among the
  impulsive transients saved. This involves improving on the current preliminary
  cuts as outlined in chapter 4, as well as writing code to precisely compare the
  lateral power distribution and polarization of candidate events to cosmic ray
  air shower models.

- Develop a calibration strategy for the cosmic ray data. I will initially use
  calibration solutions from the correlator, averaged over the observing band,
  but a better approach to calibrating band-limited impulses in the future can
  incorporate a model of the impulse response of the electronics.

- Streamline real-time commensal observations.

Longer-term goals for the OVRO-LWA

When fully commissioned, the OVRO-LWA could make precise measurements of
the three key observables– energy, X_{max} , and arrival direction– for 2000 cosmic
rays in the energy range 100 PeV to 1000 PeV in a year. This sample will improve
statistics by an order of magnitude compared to the earlier LOFAR experiment in
this energy range [30].

I would like to use a sample of 2000 cosmic rays to measure the mass composition
in the energy range 100-1000 PeV, estimating the fraction of cosmic rays in four
mass groups: hydrogen, helium, intermediate masses (carbon, nitrogen, oxygen),
and iron. A composition analysis could follow a process outlined in the flow

chart in figure 6.1. Measuring distinct mass groups offers much more information about the Galactic to extragalactic transition than does the X_{max} distribution alone, and a 2000-cosmic-ray sample would make it possible to estimate the fraction of cosmic rays in these specific mass groups with order-of-magnitude better statistics than LOFAR. These measurements may help resolve the existing tension between Auger and LOFAR. The new mass composition constraints could contribute toward pinpointing the transition energy from Galactic to extragalactic sources by clarifying (a) at which energy the main source of galactic cosmic rays stops producing light elements, (b) whether the second knee in the cosmic ray spectrum is the energy at which Galactic cosmic ray accelerators stop producing iron, (c) how abruptly the extragalactic accelerators dominate the flux, and (d) whether a second source class of Galactic accelerators is needed, and what flux of cosmic rays it must account for.

The dense antenna layout of the core of the array makes the OVRO-LWA an interesting place to test new air shower reconstruction techniques, such as the interferometric technique outlined by [87]. The current state-of-the-art forward modelling approach to X_{max} reconstruction is extremely computationally intensive. Recent simulation work [87] has proposed a much faster tomographic approach to forming three dimensional maps by back-propagating the radio signal received at each antenna. The position of peak power in these maps is sensitive to X_{max} . Simulations indicate that this technique will work best for dense arrays, such as the OVRO-LWA and the future Square Kilometer Array. Prior to the construction of the SKA, the OVRO-LWA is the only array that can test the potential of this technique.

Furthermore, if this technique is successful, this may offer an approach to simultaneous measurements of both X_{max} and shower length scale parameters, allowing hydrogen and helium air showers to be distinguished from each other on an event-by-event basis for the first time. Potentially, this independent measurement of shower length scale and X_{max} could even determine which of the existing simulations best describes hadronic strong force interactions at energies far higher than the Large Hadron Collider can probe.

Uncertainties in models of hadronic particle interactions (mainly uncertain cross sections) that cannot be probed with the Large Hadron Collider currently limit the state-of-the-art cosmic ray composition estimates [76]. However, composition-dependent differences between different hadronic interaction models affect not only X_{max} but also the shower length scale. Recent simulations [26] suggest that simultaneous reconstruction of both L and X_{max} by dense radio arrays has the potential to distinguish among hadronic interaction models and also to provide a precise measurement of the proton fraction in high-energy cosmic rays. Using the current air shower forward modelling approach, simultaneous reconstruction of L and X_{max} is far more computationally-intensive than reconstructing X_{max} alone [26]. The new tomographic reconstruction technique has the potential to open up access to simultaneous L and X_{max} measurements, potentially ruling out certain hadronic interaction models as well as certain source models. The OVRO-LWA is the array best-suited to test these techniques until the future SKA exists. If successful, this technique opens up the possibility of reconstructing more shower parameters (such as length scale in addition to X_{max}), which could distinguish between different source models as well as different hadronic interaction models.

As the first array to use radio signals alone for triggering data recording, classifying events, and reconstructing key air shower observables, the OVRO-LWA will help pave the way for future large arrays such as the SKA and also demonstrate trigger methods that may inform future array designs and tau neutrino searches.

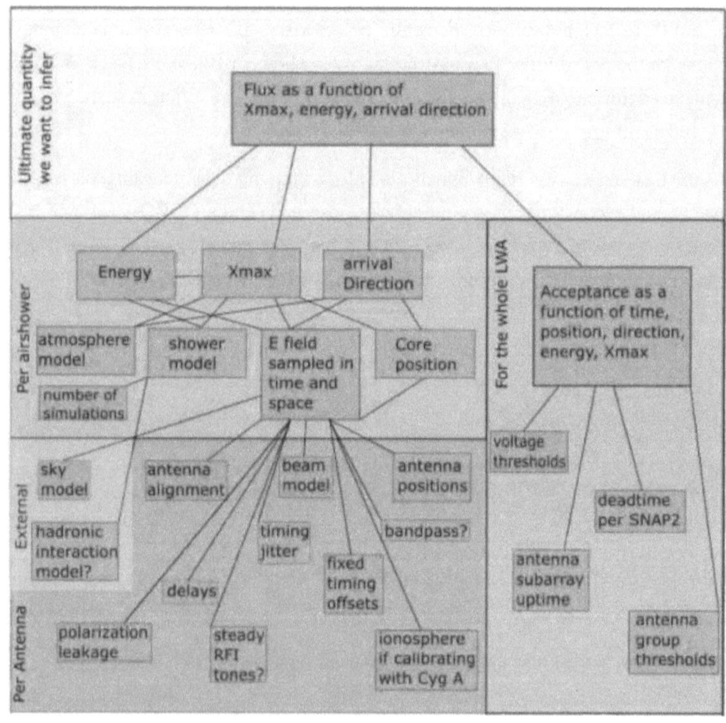

Figure 6.1: Proposed flow-chart of a composition analysis.

INITIAL SET UP OF NEW SNAP2 BOARDS

Introduction

This appendix documents the steps to set up a new SNAP2 (revision 2) (see Figure A.1), and is written as a manual for setting up additional boards. Four SNAP2 boards were set up in the lab for tests as part of the development of the digital signal processing system for the upgraded OVRO-LWA, and this manual focuses on the first set-up of those boards as an example. Basic set-up is divided into the steps listed below, which are each detailed in a subsection of the document.

Summary Steps:

1. Remove the original chassis and add fans and heatsinks.

2. Connect the board to a 12V power supply.

3. Connect to Ethernet network.

4. Connect to JTAG.

5. Load golden image with JTAG

6. Configure to be able to communicate with board via Casperfpga.

Cooling

The SNAP2 boards require fans and heat sinks for cooling. The following components are used in the lab, although at the observatory individual fans are omitted in favor of an HVAC system forcing cooling air through the whole electronics rack.

Components:

1. 45mm heatsink from Digikey[1]

---

[1] $10.44 ea https://www.digikey.com/product-detail/en/ATS-55450W-C1-R0/ATS1330-ND/1285044

Figure A.1: SNAP2 boards in the Caltech Radio Astronomy Lab. Upper left: SNAP2 board with ADC cards attached and no 40 Gb ethernet connection. Lower left: SNAP2 three SNAP2 boards with 40 Gb ethernet cables attached, and no ADC cards. Right: Lab bench with three SNAP2 boards.

2. Noctua NF-A4x20 FLX fan[2]

3. Fan Power Supply 100-240v AC to 12v DC 1A Output 3pin or 4pin PWM Connector[3]

First, I install the heatsink following its manufacturer's instructions. The plastic housing that the fan power cord plugs into on the SNAP2 differs from what the SNAP2 version 2 schematic indicates and is slightly smaller than the connector on the fan power cord, although the pins are compatible. The fan's power cord connector can be modified by shaving away a small amount of plastic, but this makes the connection somewhat unreliable. Instead, I use the external power supply described above.

---

[2] $14.95 ea. `https://www.amazon.com/gp/product/B072JK9GX6/ref=ox_sc_act_title_1?smid=A1Z5H6ZGWCMTNX&psc=1`

[3] $5.95 each `https://www.coolerguys.com/products/coolerguys-fan-power-supply-12v-1a-output -3pin-or-4pin-pwm-connector?variant=20282817937504¤cy=USD&utm_medium=product_sync&utm_source=google&utm_content=sag_organic&utm_campaign=sag_organic&gclid=Cj0KCQiAwf39BRCCARIsALXWETzAW2P-aHpk__f1zWWhVX6v8nDH_P9mun9njOPFuUqktJp0x6pkO4waAjXDEALw_wcB`

Power

The boards require a 12V power supply and draw more than 5 Amps. Each lab power supply's current limits are configured according to anticipated needs.

The SNAP2 boards have a 6-pin molex connector to deliver power to the board. Only four of the six pins are used. Apply +12V to two pins and ground the other two, as shown in Figure fig:lwa-power. Within the PCB, one of the two middle pins is shorted to a 12V pin and the other to a ground pin. **Important: There is a discrepancy between the pin voltage labelling in the SNAP2 Revision 2 data sheet and the labelling on the PCB itself. The PCB is correct.** This discrepancy and the measurements made to resolve it are detailed in a July 29 2020 Memo. Figure A.2 is the key result from that memo, repeated for quick reference here.

I use the following components for building the power cables:

1. AWG 16 wire

2. Molex Mini-fit Jr. pins: part number 0457503112

3. Molex Mini-fit Jr. connector: part number 0039012060

4. Banana connectors: part number 1325-2

Configuring the FPGA

The Ultrascale+ FPGA on the SNAP2 can be programmed via JTAG with a Xilinx USB Platform Cable[4] and Vivado's Hardware Manager software[5], or via the Casperfpga python library[6] over an Ethernet connection to the board. Using Casperfpga requires that the board is already running firwmare with a Microblaze processor and a configuration to use an Ethernet port (see section 5).

In this section I first describe programming the FPGA via JTAG. Section 4.2 describes loading firmware into memory via JTAG in order for the board to boot from this memory whenever it power-cycles, which facilitates subsequent programming

---

[4]Xilinx HW-USB-II-G $269 from Digikey `https://www.digikey.com/en/products/detail/xilinx-inc/HW-USB-II-G/1825189`

[5]`https://www.xilinx.com/support.html#documentation`

[6]`https://casper-toolflow.readthedocs.io/projects/Casperfpga/en/latest/`

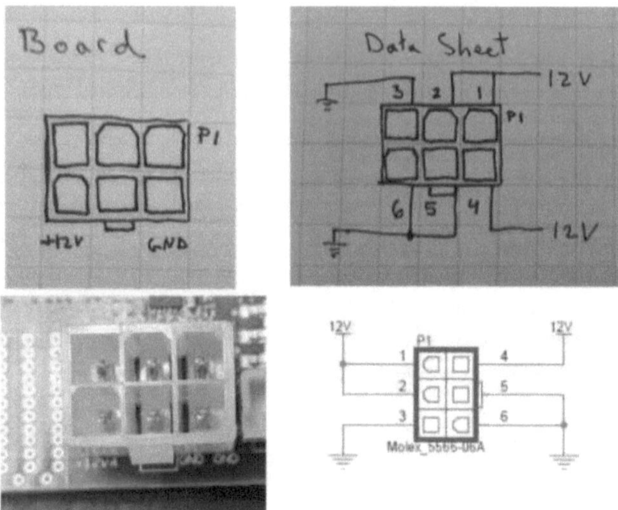

Figure A.2: The left side shows the labelling of the pins on the board, with a photograph (lower left) and a simplified diagram (upper left). The right side shows the labelling of the pins in the data sheet, with a screenshot of the data sheet in the lower right and a simplified diagram in the upper right.

without the JTAG device. Last, I describe a network configuration in order to program with Casperfpga. Casperfpga and Vivado's hardware manager have detailed documentation, so here I focus on summarizing the specific steps I did in the lab.

Programming with JTAG

1. Connect the JTAG device as shown in Figure JTAG.

2. Start Vivado in a VNC on the computer to which the JTAG device is connected. At the top of the main Vivado window, click Flow, and then open the Hardware Manager.

3. Click "Open Target" and follow the prompts to open a connection with the JTAG device (different JTAG devices will be identifiable by their serial numbers). The JTAG connection will be listed under hardware when it's open.

4. When the hardware target is open, right-click the FPGA (XCKU115_0) and select "Program Device". Then follow the prompt to choose a file to program

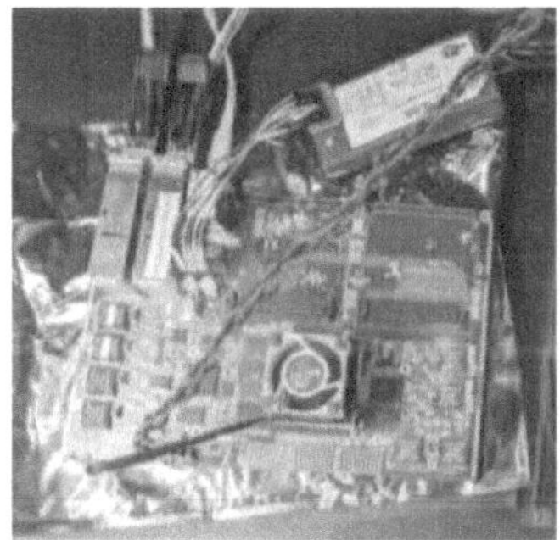 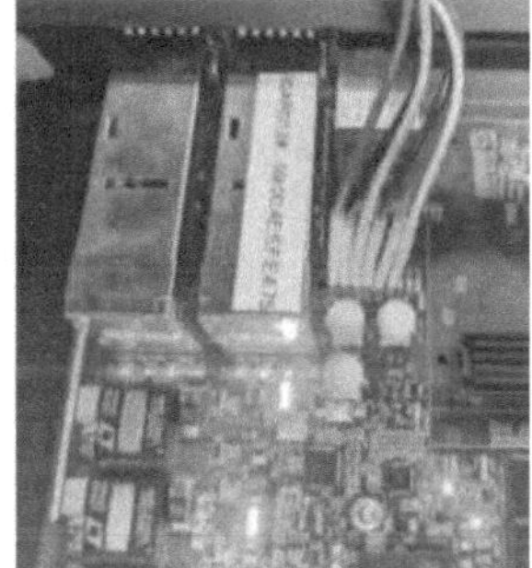

Figure A.3: Images of the JTAG device connected to a SNAP2 board.

it. In the Casper toolflow build directory, the output file that allows the FPGA to be configured via this method is myproj/myroj.runs/impl_1/top.bit.

5. When finished, right-click the appropriate JTAG in the hardware list to select the option to close it.

Storing firmware in configuration memory

These instructions will save firmware with which the FPGA will be configured whenever it reboots.

1. Right-click to open the appropriate JTAG connection (Named xilinx_tcf/Xilinx/<serial number>) from the list.

2. Right-click the FPGA [xcku115_0] under this open JTAG and select "Add Configuration Memory Device".

3. Under select memory part, scroll to micron 512Mb MT25QL512-spi-x1_x2_x4. Select it and click OK.

4. Either choose Ok when prompted to program the memory device now, or later, from the list on the left, right-click the memory part and select "Program Configuration Memory".

5. Select the appropriate .bin file (not .bit) as the "Configuration File". I use a file copied from /home/jackh/snap2v2_gold.bin on the computer Maze.

6. Wait a few minutes for the flash to program. 7. Right-click the FPGA part and hit "Boot from Configuration Memory". At this point, the FPGA should boot and request an IP within a few seconds.

Later, after programming the FPGA with other firmware (via JTAG or Casperfpga), either power-cycling the board or clicking "Boot from configuration memory" in the hardware manager will load it with this stored firmware.

Using the Ethernet port and Microblaze to configure the FPGA with Casperfpga

In order to use Casperfpga commands, the FPGA must be running a Microblaze processor and be configured to use an Ethernet interface. Compile firmware with the "use Microblaze" option selected, and include an Ethernet block in the design. Tie all the inputs of the Ethernet block to zeros of the appropriate data type terminate its outputs. Give this Ethernet block a name that is alphabetically sooner than the name of any other Ethernet blocks in the firmware. The file snap2v2_gold.bin mentioned in the previous section will configure the FPGA to use the Microblaze and seek an IP address with the LOWER of the two 1Gb ports.

Network

Connect the SNAP2's lower 1-Gigabit Ethernet port to a network with a DHCP server running. Load firmware initially by JTAG so that the SNAP2 requests an IP address after power-cycling. Flush the DHCP leases on the computer running the DHCP server to ensure the SNAP2 receives a recent address assignment:
1. Turn off the SNAP2.
2. Stop the DHCP server: sudo /etc/init.d/dnsmasq stop
3. Clear the existing leases: sudo rm /var/lib/misc/dnsmasq.leases
4. Restart DHCP: sudo /etc/init.d/dnsmasq start
5. Turn on the SNAP2.

Look in /var/lib/dnsmasq.leases or /var/log/syslog to find the MAC address and IP of the SNAP2. With this information, associate a host-name to that MAC address by editing /etc/ethers on the computer running the DHCP server. In order to set which IP address that board will always be assigned, edit /etc hosts on the DHCP server computer. After editing /etc/hosts or /etc/ethers, flush the leases again for the board to be assigned the new name and IP address.

Note that if the device is open in the Vivado hardware manager when you power-cycle the board, it won't program itself from configuration memory automatically. In this case, tell it to program from configuration memory by selecting that option within the hardware manager. If you close the target before power-cycling, then it will program itself from configuration memory automatically after powering on.

Casperfpga

To set up Casperfpga, clone Jack Hickish's caltech-lwa repository [7] and install Casperfpga as follows.

```
conda create --name py36 python=3.6
conda activate py36
cd caltech-lwa/control_sw/libs/Casperfpga/
pip install -r requirements.txt
python setup.py install
```

I have confirmed that a python 3.8 environment is compatible as well. Next, to program a SNAP2 via Casperfpga, do:

```
conda activate py36
python
import Casperfpga
fpga = Casperfpga.CasperFpga("snap-name")
fpga.upload_to_ram_and_program("path/to/firmware.fpg")
```

It will take several minutes to configure the FPGA, and TFTP warnings are normal. If it is successful, the command will return and fpga.listdev() will show the names of the software registers in the design.

A Software Environment for Compiling Firmware

Ubuntu 18.04, Matlab 2019a, and Vivado 2019.1 work with the version of the CASPER libraries adapted by Jack Hickish in the Caltech-LWA Github repository for compiling firmware for the SNAP2s for the OVRO-LWA[8]. See the CASPER

---

[7]`https://github.com/realtimeradio/casperfpga/tree/lwa352` I used this commit: eeacab5fe017bc000675d410597bba0a173214bd

[8]`https://github.com/realtimeradio/caltech-lwa`

documentation for more details, but note that these versions are slightly different from the posted compatibility matrix.

I installed Matlab 2019a using Caltech's site licensing. I installed Vivado 2019.1, using these instructions `https://www.xilinx.com/myprofile/subscriptions.html` and chose "Vivado HL System Edition", keeping the default options. I installed the Casper toolflow as follows:

```
cd /home/kplant/software
sudo apt-get install python3-venv
python3.6 -m venv /home/kplant/software/casper_venv
source /home/kplant/software/casper_venv/bin/activate
git clone https://github.com/realtimeradio/caltech-lwa.git
cd  caltech-lwa/
git submodule init
git submodule update

cd /home/kplant/software/caltech-lwa/mlib_devel
pip install -r requirements.txt
```

I set up the configuration file /home/kplant/software/caltech-lwa/startsg.local.jarvis as follows:

```
#!/bin/bash
####### Edited for compiling Snap2 firmware on Jarvis ######
export XILINX_PATH=/data/Xilinx/Vivado/2019.1
export MATLAB_PATH=/data/matlab/R2019a
export PLATFORM=lin64
#export XILINXD_LICENSE_FILE=/home/kplant/.Xilinx/trial.lic
export XILINXD_LICENSE_FILE=/data/Xilinx/Xilinx.lic
export LD_PRELOAD=\${LD_PRELOAD}:"/usr/lib/x86_64-linux-gnu/libexpat.so"
export CASPER_PYTHON_VENV_ON_START=/home/kplant/software/casper_venv
```

Then, to launch the toolflow, I do: 1. Open terminal in xfce vnc [9].

---

[9]Avoid using Gnome with Simulink. I got error messages when opening windows to configure certain Simulink blocks ("Recursion detected in the OpenFcn"). This is a documented problem using Simulink with Gnome. Switching to xfce fixed it, and KDE probably would as well.

2. source /home/kplant/software/casper_venv/bin/activate
3. cd /home/kplant/software/caltech-lwa/mlib_devel
4. ./startsg ../startsg.local.jarvis